DYNAMICS OF POPULATION ECOLOGY

Encyclopaedia of Environmental Pollution and Awareness in 21st Century Series

DYNAMICS OF
POPULATION ECOLOGY

Edited by
V.K. Prabhakar

ANMOL PUBLICATIONS PVT. LTD.
NEW DELHI - 110 002 (INDIA)

ANMOL PUBLICATIONS PVT. LTD.
4374/4B, Ansari Road, Daryaganj
New Delhi - 110 002

Dynamics of Population Ecology

First Edition, 2001

ISBN 81-261-0925-4

PRINTED IN INDIA

Published by J.L. Kumar for Anmol Publications Pvt. Ltd., New Delhi - 110 002 and Printed at Mehra Offset Press, Delhi.

Contents

PREFACE

Today, the cry of 'pollution' is heard from all the nooks and corners of the world, and pollution has become a major threat to the very existence of mankind on this earth. It is the major challenge of our times.

The pollution of various resources has gone to such an extent that we are unable to breathe fresh air and drink fresh water. On one hand the advancements of science and technology have added to the human comforts by giving us automobiles, electrical appliances, supersonic jets, space crafts, better medicines, better chemicals to control harmful insects and other pests, etc., but, on the other hand, they have given us a very serious problem to face that is pollution. Pollution can be defined as an undesirable change in the physical, chemical, or biological characteristics of the air, water, or land that can harmfully affect health, survival, or activities of humans or other living organisms.

Pollution does not have to cause physical harm, but it may merely interfere with human activities. Chemicals spewed into the air or water from an industrial plant may be harmful to humans and other organisms living near the plant. If expensive pollution controls are required, the plant may be forced to shut down.

India faces serious problems of environmental pollution. Water pollution from untreated community and industrial wastes flowing into our rivers and streams, increasing use of fertilizers and pesticides, dumping of organic wastes etc., are issues that need to be tackled on the priority. Air pollution, long regarded as the bane of industrialised nations looms large over our country.

Here, we are presenting this *Encyclopaedia of Environmental Pollution and Awareness in Twenty-first Century* in fifty volumes with a view to make people conscious about the seriousness of pollution in the present time and its remedies.

Editor

1 Population Structure

Population processes—births, deaths, and movements of individuals—express the varied interactions of organisms with their environments. Evolution by natural selection and the regulation of community structure and ecosystem function transcend the individual organism and can be comprehended only in terms of the dynamics of populations. Hence the focus of much of ecology at the population level. Individuals within populations resemble each other in any respects. Thus for many purposes one may determine average birth and death rates for small samples of individuals and multiply them by the total population size to characterise the overall dynamics of a population. But populations also are heterogeneous. Genetic variation in the interaction of individuals with their environments results in evolution. Variation in the suitability of habitat leads to parallel variation in population density and may determine the net movement of individuals within the population.

Population characteristics also vary with respect to sex, social position, and the accumulated effects of accident and chance. The population has temporal continuity because individuals alive at one time are the descendants of others alive at an earlier time. The population also has spatial continuity because he individuals in different parts of the distribution are descended from common ancestors; generally the greater the distance between two individuals, the more remote in time is their common ancestor. Individuals within a population derive their genes from a common pool and thus share a common history of adaptation to the environment. This is why it is valid to consider the traits of individuals in a population as approximately uniform, particularly compared to differences in those traits between populations. Individuals interact with one another ecologically, either directly through social contact or indirectly by utilizing shared resources. Within its lifetime, an individual affects, and is affected by, a relatively small fraction of the total population.

Each individual's sphere of ecological influence depends in part on the distance it travels. But regardless of how much individuals move within a population, the dynamics of one local population may be partially independent of those of others. When dispersal in infrequent and selective factors differ, subpopulations may also evolve independently, leading to the formation of genetically distinctive ecological races, or ecotypes. Understanding the dynamics of a population requires that one characterise both local processes affecting birth and death and the movement of individuals between localities.

In principle, the tasks of population biology are straightforward. In practice, population variables are difficult to estimate; much of this chapter is devoted to the techniques of population study. The spatial structure of a population has three main properties: distribution, dispersion, and density. The distribution of population is its geographical and ecological range, determined primarily by the presence or absence of suitable habitat conditions. Hence the distribution of the sugar maple in the United States and Canada is limited to the east by the Atlantic Ocean, to the west by low precipitation, to the north by cold winters, and to the south by hot summers.

Undoubtedly much suitable habitat exists throughout the world, especially in Europe and Asia, where sugar maple does not occur because it has not had the opportunity to disperse to such areas. Within the geographical range of a population, individuals will be fund only where the habitat is suitable. Sugar maples do no grow in marshes, on serpentine barrens, on newly formed sand dunes, in recently burned areas, or in a variety of other habitats that are simply outside their range of ecological tolerance. Hence the geographical range of the sugar maple is a patchwork of occupied and unoccupied areas, just as it is for most other species. For example, climate, topography, soil chemistry, and soil texture exert progressively finer influence on the geographical distribution of the perennial shrub *Clematis fremontii.*

Climate and perhaps interactions with ecologically related species restrict this species of Clematis to a small part of the Midwestern United States. The variety riehlii occurs only in Jefferson County, Missouri. Within its geographical range, *Clematis fremontii* is restricted to dry, rocky soils on outcroppings of dolomite. Small variations in relief and soil quality further restrict the distribution of Clematis within each dolomite glade to site with suitable conditions of moisture, nutrients, and soil structure.

Local aggregations occurring on each of these sites consist of many more or less evenly distributed individuals.The horned lark (*Eremophila alpestris*), a small songbird of short grass-lands, occurs more widely in North America than Clematis fremontii but its local distribution reflects subtle variation in habitat; even some parts of an individual's territory are utilized more than others. The distributions of populations include all the areas they occupy during their life cycle. Thus the geographical range of salmon includes not only the rivers that serve as spawning grounds but also vast areas of the sea where individuals grow to maturity before making the long migration back to their birthplace. Many birds make annual migrations to warm climates during the winter months. Thus the range of the blackburnian warbler lies entirely within the northeastern United States and southeastern Canada during the summer, but within Central America and northern South America during the winter. The size and year-to-year variations of the population of such migratory species are determined by interactions of individuals with their environments throughout the year. Within the distribution of a population, dispersion characterizes the spacing of individuals with respect to one another, forming patterns that vary from tight aggregation in discrete clumps to even, or regular, distribution. Between these extremes is the pattern of random dispersion, in which individuals are distributed throughout a homogeneous area without regard to the presence of others.

Regular and clumped distribution patterns derive from different processes. Even spacing—sometimes called overdispersion—commonly arises from direct interactions between individuals. The maintenance of a minimum distance between each individual and its nearest neighbor results in even spacing; for example, in their crowded collonies, seabirds place their nests just beyond their neighbors' each. Plants situated too close to larger neighbors often suffer the effects of shading and rot competition; as close neighbors die, spacing of individuals becomes more even. Clumping, or aggregation, results from social tendencies of individuals to form groups, the clumped distributions of resources, and the tendency of progeny to remain in the vicinity of their parents. Salamanders that prefer to live under logs exhibit clumped distributions corresponding to the patterns of fallen dead wood. Birds that travel in large flocks often aggregate to find safety in numbers.

Mathematical measures of dispersion can be compared to the values expected to random spatial distributions. They may be calculated from either the distributions of individuals among sample plots or the distances between individuals within a population. Suppose you establish a sampling area divided into plots or other units of equal area, and count the number of individuals in each of the plots. Some of the plots will have one individual, some many, and others none, depending on the density of the population and the size of the plots. Consider, for example, the number of red mites counted on each of 150 apple leaves: 70 of the leaves had no mites, 38 had one, 17 had 2, and so on; no leaf had more than 7 mites. Unless we can determine whether individuals are more clumped or more spaced than one would expect if they were randomly distributed, we can say little about this pattern other than to describe it. According to the Poisson distribution, the variance in the number of individuals per plot is equal to the mean. Hence the ratio of the variance to the mean *(V/M)* provides a convenient index to the degree of dispersion within a population. With even spacing, most plots will have the same number of individuals and the variance in number of individuals per plot will be considerably less than M (hence $V/M < 1$). With clumping, there will be many plots with no individuals and many with a large number, and the variance will considerably exceed the mean ($V/M > 1$). The variance is estimated by where *F(x)* is the frequency of plots with x individuals and *N* is the total number of plots. In the mite example, the mite example, the variance is 2.44 and the ratio of the variance to the mean is 2.1, showing clumping. Because the sampling distribution of this ratio is known, we can determine whether a given value represents a significant departure from 1, the expectation of a random distribution.

When organisms are distributed among discrete units of habitat, such as leaves on a tree, individual plants, and small ponds, it is reasonable to compare the frequency distributions of number of individuals per unit against a Poisson distribution. When individuals are distributed over a more or less uniform area, one may either arbitrarily divide the area into subplots and treat each as a sampling unit, as we did above, or use a plotless sample method in which the degree of aggregation within a population is estimated by distances between individuals.

P.J. Clark and F.C. Evans devised an index to dispersion based on the distance between nearest neighbours within a population. When individuals are distributed at random with respect to each other, the expectation of the distance between an individual and its nearest neighbor [E(r)] is one-half of the square root of π divided by the density of individuals (provided, of course, that distance and area are measured in the same units). For example, at a density of 4 individuals per square meter the expected (average) nearest-neighbour distance is 0.44 meters. The ratio of the observed average nearest neighbour distance to the expected value provides an index to dispersion. Ratios less than unity indicate clumping; those greater than unity indicate even spacing. Other nearest neighbor methods are discussed by Pielou, Diggle, and Warren and Batchlor; in Cormack and Ord.

The dynamics of populations describe changes in the number of individuals over time. In order to measure change from one period to the next, one must be able to count individuals. Rarely can one enumerate an entire population. One such case in the whooping crane, hose population numbers fewer than 200 individuals; it winters in small area of the Aransas National wildlife Refuge, where taking a census is easy. Where counting entire populations is impractical, either because of numbers of area of distribution, biologists estimate densities within defined areas as indices to population size. Density itself is an important property of a population because it indicates the potential intensity of interaction among individuals. General methods for estimating density may be found in many manuals and texts, among them Cormack, Patil, and Robson, Grieg-Smith, Mueller-Dombois and Ellenberg, Ralph and Scott, and Southwood. When individuals are not mobile-plants and sessile marine invertebrates, for example-their densities may be estimated by counting individuals within plots of known area. When individuals move between plots faster than the investigator can count the number within a plot, an alternative is the mark-recapture method. Suppose that M specially marked individuals are released into a population of size N. If the marked individuals mix thoroughly with the unmarked ones, a subsequent sample of the population should contain proportion M/N marked individuals. Therefore, in a sample of n from the population, the number of marked individuals recaptured (x) should be

$$x = nM/N$$

Because we know the quantities x, n, and M, this equation can be rearranged to solve for population size

$$N = nM/x$$

This expression is known as the Lincoln Index to population size. The standard error of N, $N\sqrt{(N-M)(N-n)/nM(N-1)]}$, indicates the accuracy of the estimate of population size. By way of illustration, suppose we capture 25 sunfish in a small pond, mark them by a small, distinctive notch in the tail fin, and then put them back into the pond. Two days later, by which time we suppose the marked fish will have mixed thoroughly with the unmarked ones, we take another sample of 35 fish, or which 14 are marked. The Lincoln Index estimate of the population size is 35 X 25/14 = 62.5 individuals with a standard error of 8.7. Because we can be 95 per cent certain that the true number of fish lies between two standard errors above and two standard errors below the estimate, the probable number of fish in the pond in between 45 and 80 individuals. Small samples do not provide accurate estimates.

Furthermore, violation of the assumptions of the model, that marked fish mix thoroughly with the population and that marked and unmarked fish are caught with equal probability, may bias the estimate, and it is difficult to ascertain the validity of these assumptions. For example, suppose that, instead of trying to estimate numbers of a confined population, like those of fish in a small pond, we wish to estimate the density of flies in an open grassland. Because the grassland poses no practical limits to the dispersal of files, the total population that the marked individuals mix with increases as the flies disperse from the point of release. Hence our estimate of the fly population would increase with time, a vary undesirable result. The mark-recapture technique has been elaborated greatly to deal with some of these problems, so that analyses are now very complex. Practical methods of marking various types of organisms are discussed in Giles and Southwood.

Dealing with populations is complicated by the fact that numbers change over time and space. A goal of population biology is to provide explanations for these patterns, but the Initial challenge is to describe the patterns themselves. There probably is no such thing as a characteristic population of a species; one's perception is inevitably colored by where ad when one looks. Records of the population of the chinch bug (Blissus leucopterus) in Illinois between 1840 and 1939 illustrate this point. These records exist because chinch bugs damage cereal crops; the Illinois State Entomologist's Office and later the State Natural History Survey Division determined the importance of monitoring the population. The populations were estimated from county reports of crop damage, which have been shown to be highly correlated with chinch bug density. To illustrate the numbers involved, during 1873, when crop damage was serious over most of the state, a ballpark estimate for the population would have been 1000 chinch bugs per square meter over an area of 300,000 square kilometers, or a total of 3 X 10 (300 trillion, more or less). In 1870 and 1875, hardly and damage was reported. Severe outbreaks were often restricted to small portions of the state, sometimes in the north, sometimes in the south. Temporal variation in population dynamics within a locality may also be perceived in the age structure of a population; that is, the relative frequencies of individuals of each age.

The age composition of samples from the Lake Erie commercial whitefish catch for the years 1945-1951 shows that during 1947, 1948, and 1949, most of the individuals caught belonged to the 1944 year class. Age can be estimate from growth rings on the scales, and the data suggest that 1944 was an excellent year for spawning and recruitment, particularly compared to the several years that followed. Variation in the annual entrance of individuals into a population has also been revealed by the age structure of stands of trees. Age may be estimated from the growth rings in the woody tissue of the trunk of the tree, one ring being added each year under normal circumstances. The pattern in a virgin stand of timber surveyed near hearts Content, Pennsylvania, in 1928, is typical and shows that the recruitment of most species was heterogeneous over the nearly 400-year span of the record. Many white pines became established between 1650 and 1710, undoubtedly following a major disturbance, perhaps associated with the serious drought and fire year of 1644. Fire can open a forest enough to permit the establishment of white pine seedlings, which do not tolerate deep shade. In contrast, the age distribution of beech, a species that can Local population changes may be due to the movement of individuals between populations. Such movement is frequently referred to as dispersal or, when discussed with respect to a particular population, as emigration and immigration. Dispersal, particularly over long distances, is difficult to measure because its detection requires the recapture of marked individuals.

Most attempts to estimate dispersal come from studies of population genetics in which investigators wish to determine the genetic structure of a population and estimate both the effective population size for evolutionary processes and the amount of gene flow between populations. The first attempts to measure dispersal n natural populations were the experiments of Theodosius dobzhanbsy and Sewall wright, who described the movement away from a release point of fruit flies (Drosophila) distinguished by a visible mutation. Traps at various distances from the release point captured some of the marked flies as they dispersed through the population. Dispersal is described by a number of mathematical indices, each making different assumptions about the structure of the population. The principal distinction is these models hinges on whether individual are continuously distributed over space, as in a population inhabiting a spatially homogeneous environment, or individuals are aggregated in discrete subpopulations, perhaps because their preferred habitat is distributed. The latter are referred to as stepping stone models.

The simplest model of dispersal is that of random movement, analogous to Brownian motion in physics. With respect to a fixed point of release, some random movements will take an individual farther away and some will bring it closer, although on average distance will tend to increase with time. The position of any individual at any moment is the sum of many random increments of distance; the probability of finding an individual at a given distance from the release site is described by the "normal" frequency distribution. This is a bell-shaped curve, whose peak coincides with the release with (that is, distance = 0) and whose breadth is characterized by a single variable, the standard deviation(s). The value of provides a convenient index to dispersal distance. A felicitous property of the standard deviation as a measure of dispersal is that the variances (s^2) of the distances add over time. Thus when s describes the distribution of dispersal distances after 1 day, the variance in distances after 2 days will be $2s^2$, and after t days it will be ts^2. Thus if adults of *D. pseudoobscura* live an average of 23 days and the average dispersal parameter(s) per day is 152 meters, (23 days) would be the square root of 23 X 151^2, or 743 meters.

Sewall Wright defined neighborhood size within a population as the number of individuals within a circle of radius $2s(t)$, where t is the average reproductive life span of the individual. Although Wright devised this parameter to describe the genetic structure of a population, it also provides an index to the number of individuals within a circle of radius $2s(t)$, where t is the average reproductive life span of the individual. Although wright devised this parameter to describe the genetic structure of a population, it also provides an index to the number of individuals in a population that are potentially coupled by strong interactions. In the case of *D. pseudoobscura*, Crumpacker and Williams estimated densities to be about 0.38 flies per 100 m^2. The area of a circle of radius $2s(t)$ is $4\pi[s(t)]^2$, which would have included an estimated 26,387 individuals.

George Barrowclough used recaptures of birds banded s nestlings to estimate s(t) for several small songbirds in the United States. Because juvenile birds usually move much farther form their birthplaces to breed than adults move from one year to the next, Barrowclough distinguished $S_{juvenile}$ and S_{adult} and combined them according to the rule of adding variances; that is, $s^2 (t+1) = s^2_{juv}, s_{ad}{}^2 t$ where t is the expected life span of an adult bird (the juvenile stage lasts about 1 year, from hatching to first breeding). For 8 species $s(t+1)$ varied between 344 and 1681 meters, densities varied between 16 and 480 individuals per square kilometer, and neighbourhood size varied between 151 and 7679 individuals. Notice that the neighbourhood sizes of these slowly dispersing snails are on the same order as those of small birds. From mark-recapture data gathered by Frank Blair on the rusty lizard (Sceloporus olivaceous) near Austin, Texas, H.W. Kerster estimated s(t) to be 89 meters and neighbourhood size between 225 and 270 individuals.

In most studies of dispersal, the principal source of information has been the movement of individuals away from a point of release. Two other kinds of observations are pertinent, however. The first of these concerns the spread of introduced populations, which can only occur by the movement of individuals. The second concerns the genetic differentiation of local populations, which is hampered by movements of individuals. The European starling spread almost 2500 miles across the United States in 60 years, an average rate of about 40 miles per year. Adult starlings tend to nest in the same area year after year and so most dispersal is accomplished by young birds. Furthermore, the westward spread of the starling was characterized by frequent sightings of nonbreeders before breeding populations were established in an area. It is unlikely that such long-distance movements of juveniles could be detected within an established population, particularly because of the limited area of intensive population study. Whereas a few long-distance movements may be sufficient to expand the border of a population, they may have less effect on the dynamics of widely separated, but established subpopulations. Given the propensity of populations for increase, a small number of colonizing individuals can grow to a large population in a short period. But those same individuals immigrating to an established population may have a negligible impact. Examples of local genetic differentiation of populations suggest that the movement of individuals between populations is small compared to the forces of selection, mutation, and random change (genetic drift). Strong selection can lead to divergence between populations in the face of considerable gene flow, as we have seen in the establishment of distinct ecotypes of plants in different habitats. The presence of apparently nonadaptive differences between close populations provide stronger evidence of limited gene flow. Such a situation occurs in populations of the small annual plant Linanthus parryae. In one small area in southern California, blue flowers-the normal flower color is white-are haphazardly distributed without apparent relation to environmental factors. Some samples of individuals separated by less than 1 kilometer differed in flower color by 100 per cent. When populations can maintain such genetic distinction, their dynamics cannot be closely linked.

The accurate projection of population growth requires a knowledge of the age structure of the population and the survival and fecundity of individuals of each age. These variables are determined in part by genotype, sex, social rank, and the effects of chance accident. Because these factors are difficult to measure, however, most models of population growth assume that individuals are uniform with respect to all traits except age. As a result, ecologists have devoted much effort to determining the age structure of populations and the age-specific birth rates and probabilities of death of individuals. These statistics collectively are known as the life table. Demographic records have been kept for the human populations of some countries for centuries, well before the development of population models that could use the data. With the introduction of such models during the 1920s, life tables became a popular tool for both laboratory and field biologists. The first life table assembled for a nonhuman species was that of Raymond Pearl and S. L. Parker for a laboratory population of the fruit fly *Drosophila meloanogaster.*

The life table sets out the fecundities and probabilities of survival for each age class of individuals in a population. Because of the difficulties of ascertaining paternity in many species, life tables usually are based entirely upon females. For some populations with highly skewed sex ratios or unusual mating systems, this can pose difficulties, but in most cases a female based life table provides a workable population model. Age is designated in a life table by the symbol x, and age-specific variables are indicated by the subscript x. When reproduction occurs during a brief breeding season each year, each age class is composed of a discrete group of individuals born at approximately the same time. When reproduction is continuous, as it is in the human population, each age class x is designated arbitrarily as comprising individuals between ages x - ½ and x + ½.

The fecundity of females, expressed in terms of female offspring produced per breeding season or age interval, is designated by b_x. (Think of b for births; m [for maternity] is used in some treatments, but this can be confused with m used here for mortality.)

The estimation of survival schedules

Survivorship can be estimated from four kinds of information: (1) survival of individuals to a particular age, (2) survival of individuals in each age class from one time period to the next, (3) ages at death within a population, and (4) the age structure of a population. The first kind of data forms the basis of the "cohort life table" or "dynamic life table," which follows the fate of a group of individuals born at the same time from birth to the death of the last individual. The life table for *Poa annua* presented is of this type. This method is most readily applied to populations of plants and sessile animals in which marked individuals can be continually resampled over the course of their life spans. Herein lies one of the disadvantages of this method: it can take a long time to collect the data, particularly if one studies redwood trees. It is also difficult to apply to highly mobile animals.

The second method, the "time-specific" life table, sidesteps the time problem by considering the survival of individuals of known age during a single time interval; say, 1 year. Thus each age-specific survival value is estimated independently for each subset of the population during the same period (hence the "time-specific" label). A limitation of this method is that one must know the ages of individuals, so it is practical only when age can be estimated by growth rings, tooth wear, or other reliable index. The distribution of ages at death was first used to construct a life table for a nonhuman population by Edward Deevey, who worked from Olaus Murie's data on the ages of death of Dall mountain sheep in Mount McKinley (now Denali) National Park, Alaska. Murie estimated age at death from the size of the horns, which grow continuously during the lifetime of the sheep. In all, the found 608 skeletal remains; 121 sheep were judged to have been less than 1 year old at death; 7 between 1 and 2 years, 8 between 2 and 3 years, and so on, Deevey reasoned that all 608 ded sheep must have been alive at birth; all but 121 that died during the first year must have been alive at the age of 1 year (608 - 121 = 487), all but 128 (121 dying during the first year and 7 dying during the second) must have been alive at the end of the second year (608 - 128 = 480), and so on. He built his life table in this fashion until the oldest individuals had died, during their fourteenth year.

Deevey's age-at-death method is frought with potential bises. In particular, it assumes that an equal number of newborn form the basis for each age class of deaths. This assumption is violated in expanding and declining populations, in which younger and older age classes, respectively, are over-represented. Life tables constructed from the age structure of a population at a particular time suffer the same problem. The number of individuals of age x alive at time i is simply the number of newborn x years in the past times the survivorship to age x. Hence survivorship can be estimated by $l_x = n_x(i)/n(i\text{-}x)$. If one is willing to live dangerously and assume that equal numbers of offspring are born each year, then $n_0(i - x) = n_0(i)$ and $l_x = n_x(i)/n_0(i)$. The life table provides the foundation upon which all population models are built. But the particular form of a model depends on two aspects of the age structure of a population, which, taken together, result in four possible combinations. First, reproduction may be either discrete or continuous. Second, generations may overlap or not. The dynamics of each type of population are modeled in a slightly different way.

2 SOCIAL BEHAVIOUR

The behavioural phenomena are wide-spread throughout the animal kingdom, and have considerable significance in population ecology. They are territorialism and dominance hierarchies.

— *Territorialism* is essentially a social pattern of spatial utilization by which individual or groups have control over certain units of space.

— *Dominance hierarchies* are an ordered systems which determine individual access and priorities to natural resources.

Both of these phenomena relate to population ecology intimately because they represent systems of behavioural control which affect the abundance and distribution of animals, their reproductive patterns and their mortality patterns.

Many studies of territorial or rank oriented species have shown for example, that individuals without territory and low ranking individuals have less reproductive success and higher mortality rates than those who hold territory and/or have higher rank. A territory has been defined as any area which is defended against other member of the same species. It may be occupied and defended by an individual as in the case of a male stickleback fish (*Gasterosteus aculeatus*); by a pair as in the case of many birds or by a social group, as in the case of vervet monkeys (*Cercopithecus aethiops*) and gibbons (*Hylobates lar*). It may encompass most or all of the occupant's home range including his nesting and foraging areas, as in primate territories; it may include only a limited area around a nest site of feeding site, as in many birds; or it may involve only a very small plot of ground on which mating occurs, as in the Uganda kob (*Adenota kob*), an African antelope. Hence, territorialism is a highly variable and complex phenomenon, and involves a wide spectrum of behaviour patterns.

In all cases, however the essential feature is that individual animals, or groups of animals, have controlling ownership over a given piece of property. Within this property they have the right of utilization over the space and the resources contained therein. Territories are established and maintained by a great variety of behaviour patterns and communicative displays. Initially, the establishment of a territory may involve active aggressive behaviour and fighting. The territorial combats of siamese fighting fish, jungle fowl, elephant seals and wild antelope are well known. Those combats are often vigorous and are true tests of strength, injuries are common, and in rare cases, death may occur. Most frequently, however, the vanquished animal merely retreats and vacates the territory. Once established territories are most often maintained by communicative signals and displays. These signals may be visual, auditory or olfactory in nature. Most fish have visual territorial displays which consist of spreading the fins and brightening the colors.

Some fish have active sound production in territorial calls, as for example the Atlantic toadfish (*Opsanus tau*). Most birds which are territorial have a combination of vocalizations and visual displays. The male song bird of many species perches in a conspicuous spot and repeatedly gives forth his song, which serves more often as territorial threat than as a mating attraction. Mammals typically use a combination of auditory and olfactory signals as communicative displays in the maintenance of territory. Some of the arboreal primates, such as the gibbon and the howler monkey (*Alouatta palliata*) have an advanced vocal pattern which conveys information about the position of the group and indicates that the territory is occupied.

Many carnivores, such as the wolf, lion and tiger (*Felis tigris*) have territorial calls which can be heard for several miles, and they also have olfactory signals in scent glands and urine which are used to mark the boundaries of territories. The scent signals may persist for days or even weeks, and thus serve as more stable cues than serve as more stable cues than vocalizations which fade rapidly. Territorialism is common among many higher invertebrates (crustacea, insects, spiders, etc.) and throughout all vertebrate classes. Territorialism can be found in fish amphibia, reptiles, birds and mammals, which seem to operate on much the same principles and have basically the same functions.

Although this is true, it must not obscure the fact that there are many animals which do not show territorialism. Many species do not stake out specific territorial claims and do not actively defend given tracts of habitat. Often one can find closely related species which are territorial in one case, and nonterritorial in another. In ungulates, for example, the Indian blackbuck (*Antilope cervicapra*) is territorial, whereas the Indian swamp deer (*Cervus duvacelli*) is not. In primates, the gibbon and the howler monkey are clearly territorial, whereas the rhesus monkey (*Macaca mulatta*) and the chimpanzee (*Pan troglodytes*) are not. Even within the same species there are circumstances in which the individuals of some populations are territorial distributed and other circumstance in which they are not. This is true in ungulates for example, many of which are territorial during the rut or mating season, but the same individuals are non-territorial at other times of the year.

Apart from this obvious correlation with breeding, however, there is evidence in some species that territorialism is related to the population density and/or environmental circumstances. In the Japanese salmon (*Plecoglossus altiveli*), individual fish are territorial at low population density, but as density increases, individuals give up their territories and form a school. Wild house mice (*Mus musculus*) show similar behaviour with territorialism at low population densities, and mass group behaviour at high population densities. This variability is related to environmental factors and differences in social behaviour of key individuals. In view of these facts, it does not seem warranted to consider territorialism an instinctive pattern of behaviour. Territorialism is certainly common and wide spread, but by no means universal within limited animal groups or even within a single species.

Dominance Hierarchies

Whereas territories function to allot resources on a spatial basis, dominance hierarchies allot resources on an individual priority basis within the same physical space. Hierarchical systems have been observed in all vertebrate classes and in many invertebrates. Dominant animals have priority access to food, mates, nest sites and resting locations. Dominance is often manifested between animals by single physical displacement by the displacement of one individual by another at a feeding or resting site. Occasionally it involves displays or threats, and rarely it may involve direct fighting. The latter is more common in initial encounters in which the dominance rank or social status of each individual is not yet determined, but once established the hierarchy is then usually established by display or by social memory. Once a rhesus monkey has established high rank, it does not have to constantly display this rank in a stable social group. The high rank of some individuals and low rank of other become an accepted social norm within the group. Only when the situation becomes unstable through death, emigration or invasion does there have to be an avert re-establishment of rank within a dominance hierarchy. So far as we have seen territorialism and dominance hierarchies as two types social systems which play a regulatory role in population dynamics, the first by regulating of space second by controlling individual priorities within a common space.

In some animal species and populations, territorialism and dominance hierarchies apparently serve as sensitive feedback mechanisms which influence dispersal, mortality reproductive success and population size. In other species and circumstances, territorialism and hierarchical structures are often subject to breakdown and collapse. Ecological consequences are then profound and extensive. In many rodent populations, excessive population growth and crowding is often associated with the following phenomena: a breakdown of stable territoria-lism and dominance structure, reduced fertility as evidence by lowered pregnancy rates, high prenatal loss and low infant survival, poor nest construction and poor maintenance, nest fouling and pollution, high aggression and wounding deteriorating health in the majority of the population, abnormal aggregations of individuals in the 'behavioural sink' pattern and finally population decline.

Social behaviour

The detailed study to territorilism and dominance hierachies is really in the realm of ethology and animal behaviour, but the functions and consequences of these behavioural systems have a great deal of ecological significance. It has been abundantly demonstrated in many territorial animals that the holders of territory have the greatest breeding success. In fact, in some animals, non-territorial animals usually fail to breed. Similarly in those social groups characterized by dominance hierarchies the high ranking animals have the greatest breeding success and the low ranking members have less success or fail entirely in breeding efforts. In chicken flocks, low ranking roosters show frequent counting behaviour, sometimes more than dominant rooster, but they have relatively little success in completing copulations. It thus seems clear that the social systems of territorialism and dominance hierarchies have a limiting function in population dynamics; that is, they limit the breeding to those individuals which are ecologically and behaviorally most successful in establishing territory or high rank.

Theoretically, as population densities increase, and space becomes more limited, a smaller proportion of individuals within the population is able to achieve territories or high rank, and thus total reproduction is reduced in relation to population size. Ideally, these behaviors are density-dependent population limiting factors; that is,.the higher the density, the greater the limiting effect on the population. The role of social behavior in population regulations has been discussed at great length by Wynne-Edwards.

Wynne-Edwards proposed that the basic long-term population level of many animals is established by food and resource availability, but that social behaviour establishes a more immediate regulating influence on population size and density. This influence on social behaviour, he felt, comes into play before critical population levels are achieved; that is, before starvation and/or damaging violent aggression are conspicuous.

Thus, according to Wynne-Edwards most animals have intrinsic, self-regulatory population mechanisms which operate via social and behavioural feedback systems. These social and behavioral feedbacks are usually ritualized displays which relate to territorialism and dominance hierarchies, and thus convey information about population density. Therefore, regulatory feedback is provided according to the intensity of social competition. In other words, ritualized displays, or conventionalized rivalries, regulated breeding success and thus population numbers. Wynne-Edwards provides numerous examples of this hypothesis, including the complex display patterns of gallinaceous birds, (prairie chickens, sharp-tailed grouse, peacocks, bower birds et al.) waterfowl and many song birds. He also considers the vocal choruses of amphibians, the schooling of fish, and the flocking and birds and mammals as examples of complex social behaviour which serve the function of conveying information about population density and social competition.

Many aspects of Wynne-Edwards' theory are controversial especially when he theorized that most social behaviour has actually evolved as a means of controlling population levels. Wynne-Edwards proposed, for example, that territorialism has arisen as a social means of preventing over population. This theory has met substantial criticism from other ecologists and students of behaviour. These authors felt that there is little reliable evidence to support this hypothesis even though it is logically very attractive. Brown pointed out that, "Population regulation is never completely under the control of the species by itself, but depends in a complex way on interactions between members of the ecological community." Thus, Brown felt one must consider the interaction of all factors, including predation, disease, food supply, interspecific dynamics of any given species.

Brown found satisfactory evidence that territorialism tends to space individuals over available habitat, that it increases emigration and morality rates in individuals unsuccessful in finding territories, and that it has arisen in response to individual selection resulting from aggressive competition. The more detailed aspects of this controversy are out side the scope of introductory ecology. On the more basic parts of the hypotheses—namely, that social behaviour provides some regulatory feedback on population ecology—there is certainly no question that this is true in most animal populations. Even this concession must not obscure the fact, however, that some animals have such poorly developed and/or variable responses to increased density and crowding that they can readily breed themselves into starvation or conditions of violent aggression and extreme social strife. White-tailed deer populations, for example have frequently increased to the point of extensive starvation. Lemming and other rodent populations have also increased to the point of mass behavioral strife where social mortality becomes extensive.

Quite probably, some human populations are also in this category, providing example of populations in which social feedback limiting population growth often does not into operation until the point of starvation arrives (as in China and India), or until the point of violent and destruction aggression breaks forth (as in many wards in which population pressure has had a causative influence) or until social morality becomes limiting (as in countries where abortion is a significant form of population regulation). It is scientifically important to consider animal population in which environmental crises and/or behavioural breakdowns occur. In terms of human value judgments these seem to be cases of population regulation by wasteful and sometimes tragic means. In evolutionary terms, however, many of the populations showing these phenomena are of highly exploitative natures, meet new environmentral situations had expand into available opportunities. Perhaps their lack of frigid, stereotyped behaviour and ecology, and their lack of precisely tuned population feedback systems, enables them to seize each new ecologic opportunity to the fullest extent. Some of the best examples of such species are the domestic rodents especially the wild house mouse, the wild Norway rat (*Pattus norvegious*) and the Asian bandicoot rat (*Bandicota bengalensis*). Numerous ecological studies over the last 25 years have shown these species to have the following population characteristics:

1. They have tremendous reproductive potential and under certain environmental circumstances they often show invasive and eruptive population behaviour.

2. Populations can readily exceed the carrying capacity of the environment and this can result in sudden population decline.

3. Populations fluctuate widely, natural control is often abrupt and "ecologically wasteful", in terms of individual health and welfare.

4. Population regulation often involves social disruption and destructive behaviour.

These characteristics have been observed frequently in both natural and artificial populations of rodents over the past 40 years. The natural social behaviour of the Norway rat and wild house mouse includes both territorial and hierarchical behaviour. Females tend to the territorial around the nest site so that they normally exclude adult males and strange females. Often two of three females nest together in a communal agreement. Males are also territorial if population density is not excessively high, and high ranking males defend a system of runway and the females living therein. The size and patterning of the territories depend upon the rank of the male, the density and social structure of the population and the nature of the habitat. In a stable habitat both population levels and social structures of domestic rats might remain quite stable over a period of time.

But the habitats of domestic rodents are characteristically unstable with changing amounts of food (around farm, for example where harvests and food stores come and go), changing mortality patterns (as control compaigns) or disease outbreaks wax and wane) or changing cover conditions (as trash availability rises and falls) so that the populations are often experiencing drastically changing resource conditions. Under these circum-stances, the populations respond rapidly with surges or declines of growth. Under periods of rapid population growth leading to high densities, a variety of phenomena have been observed. There is often rapidly increasing social contact resulting in an increase in aggressive interactions.

This frequently lead to a breakdown of territorial boundaries, increased crowding of the nest sites and sharply increased mortality of young nests become trampled and destroyed. Females often desert the nests or actually kill the young. Wounds become frequent, especially an adult males forming sites for mites to invade the skin, so that acarine dermatits may achieve epidemic proportions. With a breakdown of territorial behaviour, bizarre aggregations developed so that clusters of individuals all in poor physical states appeared. Calhoun referred to such aggregations as "behavioral sinks". They are self-stimulating aggregations of individuals often collectively showing abnormal behaviour patterns. Some individuals become inactive and socially withdrawn.

As these conditions progress only limited portions of the environment are utilized but these become excessively polluted. Thus, the stage is set for both high social mortality and high disease mortality and population often collapses as rapidly as it built up. After the crash if the environment recovers the surviving individuals recuperate and population growth may begin again. The detailed of this theme differ considerably in different studies and different species. In come circumstances, there may be a cessation of production so that no young are born into the crowded population, as occurred in house mouse populations studied by Crowcroft and Row.

In other circumstances reproduction may continue, but there may be massive infant mortality of all young born. In still other cases, crowding may stimulate excessive migration and disposal as in lemmings and these mass migrations are often accompanied or determinated by high mortality. In all cases, a similar syndrome prevails. There is usually a drastic change in social organization and behaviour which most often have a major influence on reproduction and mortality. In domestic rodents these changes characteristically involve a breakdown in social stability and a disruption of normal territorial, reproductive and parental behaviour. Frequently parinatal processes are most dramatically affected, so that this pre-and postnatal survival of the young is reduced.

Population Interaction

All population of living organism exist in a network of interactions with other population. Many of these interactions are subtle and complex than direct food relationships. Others are competitive or limiting to the interacting populations. The degree of intimacy in the relationships of the organism inhabiting a particular habitat varies greatly. These association may be either beneficial or harmful to both the patterns or beneficial to one partner and harmful to the other. If the relationship is beneficial either to one or both partners, it is called 'Positive Interaction' or cooperation. This may be interaspecific (Ex. Social life of inspects) or Inter specific (Ex: Commensalism, Mutualism & Symbiosis). The interrelations between individuals that are harmful to one or more of the participants is called Negative interaction. It includes "Parasitism, Predation and Competition".

Commensalims

The term 'Commensalism' was first proposed by Van-Benedan to the animals that share the same food. "Two or more animals that live together and do not enter into any kind of Physiological Union are referred to as `Commensals' and this association is called as `Commensalism! The relation is unequal in that only one individual gets the benefit from the association while the other is neither benefited nor injured. The Chief benefits from this association are (i) Shelter, (ii) Anchorage (iii) Transportation and (iv) The food supply.

Types of commensals

A commensal which attaches or lives upon the host's body is called as 'ectocommensal'. If a commensal lives within the body of the host in the respiratory or alimentary tract or in any other body cavity which opens to outside, is referred to as endocommensal. Vorticella, a Protozoan animal is an ecto-commensal which is attached to the legs of the freshwater shrimp (prawn like) by slender stalks. The 'Sucker fish' is another ectocommensal which attaches itself to the body of shark, or sword fish or seaturtle with the help of its sucker. All the ectocommensals are small in size and not burdensome to the host. The harmless protozoans in the intenstinal tract of mammals and micro-organisms in the canal system of sponges are the examples of an ectocommensals!. An example of commensalism is the remora-shark relationship where the remora fish (Echenesis naucrates) attaches to the skin of the shark by means of a strange sucker disc and transported widely and rapidly by the shark's motive power. The remora also consumes food remnants cast off from the jaws of the shark. Thus the remora benefits in two ways from this attachment and the shark is relatively unaffected although its swimming speed may be slightly impeded.

Many large organisms provide commensal harborage for other organisms externally and internally. Large tropical trees provide habits for numerous commensal plants and animals. Trunks and branches provide attachment sites for epiphytes such as orchids, and recesses between buttress roots provide attachment harborage for bats tree frogs, lizards, insects and many other organisms.

Whales provide attachment sites for barnacles, algae, and other sessile marine forms. These plants and animals are not parasitic, that is, they do not extract nutrients from the habitat provided by the host. Most animals, including man, also contain internal commensals. The human digestive tract is inhabited by several types of bacteria and protozoa such as Entamobea cole, which are neither parasitic not pathogenic, but merely residual in the alimentary canal. Another common type of commensalism occurs when various species of plants and animals use burrows or nests constructed by another species. Termite nests provide ecologic niches for more than 100 species of other animals including ants, aphids, beetles millipedes and isopod crustaceans. Tube-dwelling annelids of the genus Chaetopterus which inhabit the tidal zone of the seashore provide habitats for small crabs of the genus Polyonyx. The crab benefits by the protection of the tube, and it apparently does no harm to the annelid worm which originally built the tube. Commensalism may be obligatory, in which one organism depends entirely on another for its habitat or it may be facultative in which each organism may be independent but one is enhanced in the presence of the other. An example of obligatory commensalism is the relationship between the algae, Basicladia, and certain freshwater turtles. The algae grows only on the backs of the turtle. Similarly the bivalve mollusc Ostrea grows almost exclusively on the roots of the red mangrove off the coast of Florida. Attached to the shell of the horse-shoe crab, Limulus, are several species of molluscs, barnacles and annelid worms which again are not parasitic but only residential and some of them live only on the Limulus shell. The oyster crab is a small crustacean which lives only in the mantle cavity is a small crustacean which lives only in the mantle cavity of oysters. These all represent cases in which one species of plant or animals has evolved a dependent relationship does not involve feeding upon the tissues of the host. The possibility exists that this is one evolutionary route toward parasitism and the distinction between obligatory commensalism and parasitism is often difficult to draw.

Facultative commensalism is represented by the relationship between the borrowing owl (Speotyto cuniculario) and the prairie dog (Cynomys sp.) The burrowing owl often nests in prairie dog burrows but is not confined to such burrows. In the north polar regions, the arctic fox Clo pex lagopus) feeds in the winter time on the remains of seals killed by polar bears (Thalarctos maritimus) though it may have other sources of food as well. Probably the majority of commensal relationship are facultative. It is especially interesting to consider animal populations which form facultative commensal relationships with men. Among vertebrates perhaps the best examples are domestic rodents such as house mice and Norway rats which have increased in abundance and distribution by their association with human habitation and agriculture. The house sparrow, pigeon and Asian house crow and Asian house shrew or musk shrew have also greatly increased their population sizes through human associations. Cities, towns and farms provide increased food supplies and nesting sites of these animals. The rhesus monkey in India is one of the few primates which has become a facultative commensal of man. The rhesus can readily survive in forests without man, but it becomes especially abundant around villages and towns of the Gangetic basin where it could live on agricultural produce.

Some more examples among the inverterbrates are as follows:—

(1) The peacrabs are found living in the mouth cavity of bivalves. They steal food collected by the host and do not cause any injury.

(2) A number of micro organism live as commensals in the canal system of sponges.

(3) Polynoid polychaetes are found in the pallial chambers of gastropods and in the ambulacral groves of star fishes.

(4) A host of commensals such as worms, Brittlestars, molluscs, little fishes, prawns-crabs etc. are found among the coral reefs.

(5) Small pelagic fishes live under the umbrellar shelter of large jelly fishes.

(6) Some fishes live in the borrows of worms and crustaceans.

(7) A number of copepeds live within the Ascidiann or Sea anemones.

(8) Barnacles attach on the skin of whales or back of turtles.

In all these associations one of the partners which is generally smaller derives benefits of anchorage, shelter, food and transportation.

Mutualism

Mutualism is an association between two or more animals in which all derive mutual benefit with no harm to any of the partners. "Symbiosis" term also often used to this type of relationship (Properly refers to the intimate associations to two or more dissimilar organisms). Mutualism may be seen between animals and plants and between animals and other animals.

Ist Type : Mutualism between Plants and Animals

This association is important to the reproduction of many kinds of plant species. Plants attract insects by their sweet, nectar or colour. The insects visit plants to get their food. The plants dust them with pollen and the insects carry the pollen from flower to flower. In the distribution of seeds, plants take the help of animals. Plants provide fruits and seeds pass in their hard coat, undigested through the alimentary canal of the animal and are distributed in its droppings. For instance a species of Alacia grows little pearshaped masses of food for the ants, and hollow thrones (*Myrmecolphilous stipules*) in which the ants take shelter. In turn the ants defend the plant from the attack of other insects.

2nd Type: Mutualism between Animals and other Animals

Birds display this type of relationship. The cow bird and Bison, the Oxpecker and the antelope, crow and the cattle exhibit mutualistic relationship. In all these associations the birds render considerable service to their hosts by ridding them of ticks and external parasites and also serves as sentinels. In turn the birds receive a constant food supply. An interesting example is between a 'Crocodile' and the bird (Pluvianus aegyptius). The bird removes leeches and fragments of food in decaying condition of the crocodile. The crocodile opens its jaws and permits the search then finally the bird receives the food from the crocodile. The common starling bird may also perform the same function from the sheep. The association of ostriches and Zaebras is also said to be of mutual benefit.

Birds and Insects

Birds have also developed relationship with insects for mutual benefit. The horn bill of Africa harbours a number of insects in its nest. The female horn bill will be confined to the nest at the time of egg laying and breeding. The male bird feeds her with fruits. The most common Commensals among marine forms are the Hydroids and Sea anemones found on the shell of Hermit crab. In this association both of them are benefited. The Indian Rock (Mincus inermic services as a moving home to zoophytes small polyps are encrusted on this fish. In this relationship the fishes obtain protection from predators and they pay for it by luring large fishes within the range of tentacles. The relationship between *Hermit Crab* and *Sea anemone* is very interesting. Here the Crab obtains protection from predatory fish by the shrinking power of sea anemone. The sea anemone also gets an advantage of transportation of feeding grounds and gets food fragments when the crab captures and eats another animal.

Sponges growing on the shell of Hermit Crab protect it by producing bad odour. Here the advantages to the sponge is that it is carried to new waters for its own food. Thus, the mutualistic relationships between plants and animals, and animals and other animals prove to be beneficial mutually and reciprocally to the participants. Like commensalism, mutualism may be obligatory or faculative. An example of obligatory mutualism is provided by the relationship between the flagellate protozoan Trichonympha, and woodeating termites. These species cannot live without each other. The protozoan lives only in the digestive tract of the termite. The stomach and caecum of the horse contains millions of ciliate protozoa and bacteria which digest cellulose for the horse and provide 20 per cent of its nitrogen requirement per day. These intestinal microorganisms are essential for the normal growth and health of the horse.

An elaborate form of obligatory mutualism occurs between some ants (e.g., Cremastogaster lineolata) and aphids or plant lice (e.g. Aphis caliginosa). The ants maintain the aphids in specially constructed nests, and they feed upon a nutrient liquid secreted by the aphids. The aphids have sometimes been called "cows" because they release the secretion when stroked by the antennae of the ants. Thus the ants receive food from ants. Many other anthropods, including mites, springtails, flies, wasps and beetles, have developed similar complex relationships are based on interspecific communication systems through which the different species jointly utilize food, shelter of other resources.

Many mutualistic relationships are facultative; that is, both populations can exist without the other, but both are favoured when living together. Baboons and impalas, for example, can survive well without each other, but each provides a unique protective warning system for the other when they live in association. Squirrels facilitate the extension and propagation of hickory trees by buying the nuts, but the hickory trees can propogate without squirrels and the squirrels can likewise survive without hickory nuts, if an adequate supply of other food is available.

Man has capitalized on mutualism as a biological principle in the development of agriculture. The domestication of plants and animals originated in facultative commensalism or mutualism. At the down of agriculture, 10,000 years ago, man could live without domesticated plants and animals, but as he developed these relationships then evolved into obligatory commensalism and mutualism. Modern man obviously cannot live without the domesticated plants and animals of agriculture, and many of these domesticated forms cannot live without the special husbandry provided by man.

Symbiosis

The term Symbiosis was first coined by DeBray in 1879. It means "living together.""Symbiosis is an association between two or more organisms that live together in close physiological union for the mutual benefit." Here the term Symbiosis refers to the intimate association of two or more dissimilar organism. One of the partners may be known as Symbiont. Symbiosis offers from commensalism and mutualism in two important respects.

(1) The relation between two partner is much more intimate in Symbiosis than in other relations.

(2) It is mostly obligatory.

Symbiotic associations are known to occur between plants and other plants, animals and plants; and animals and other organisms.

First Type Symbiotic Association among Plants

This type of associations is demonstrated in (i) Fungi and algue which form lichens; (ii) Fungal mycorrhizae and the roots of many flowering plants and (iii) Nitrogen fixing bacteria and the roots of legumes. In all these associations, there is metabolic interdependence. In the lichens, algae produces food and fungi provide mineral and protection. The intimate association between the roots of flowering plants and mycorrhizae results in the formation of a composite fungus-roots structure which differs from the roots without mycorrhizae. Here the fungi require soluble carbohydrates and substances and they get them from root secretions. Here also the association is intimate the benefits are mutual. The importance of the Nitrogen-fixing bacteria

Second Type: Symbiotic-Associations Between Plants and Animals

The most notable symbiotic association between plants and animals occur between freshwater coelenterate (Chlorohydra Viridissina) and an algae. In this association the animal obtains Carbohydrates and a supply of Oxygen evolved as a result of Photosynthesis by the algae. The large in turn receive protection and Dioxide for photosynthesis released by the animal during respiration.

Hydra and Algae relationship is also one of the good examples. The mutual benefits of hydra an algae are

Chlorachydra (*animal*)	*Zoochlorellae* (*Plant*)
i) It gives phosphorus and nitrogen compounds for protein formation.	i) Plants gives Oxygen
ii) It gives Carbon dioxide for the photosynthesis in plants.	ii) Excretory materials.
iii) It gives protection to algae (Plant)	iii) Carbon dioxide (from respiration)

Planeria and an animal algae and a plant symbiotic association is also a good example. Here the symbiotic benefits are same as in coelenterate and algae. Here, the presence of algae is indispensable for the worm and this is obligatory.

Association Between Fungus and Insects

The wood boring beetles have an elaborate apparatus on the head for carrying the Fungus. Others carry Fungus in the midgust to be regurgitated for cultivation. The females of this species have an apparatus connected with evipositor which coat their eggs with fungal spores as they are laid. In all these associations, the Fungus provide food and shelter, and it grows on the excretory material of its host.

Third Type: Symbiotic Association Between Animals (Metazoa) and Microorganisms

Symbiotic associations are also seen between Metazoa and Micro-organisms many Wood-eating-insects especially termites harbour numerous protozoans in their alimentary canal. It has been shown that the gut of termites can live on a nitrogen-free diet so long as protozoa are present. This suggests that they may be capable of nitrogen fixation of the synthesis of protein from carbohydrates. Here the protozoans produce specific enzymes "Cellulose," and "Cellobiase," which are responsible for cellulose digestion. Similar association is seen in the gut of Cockroach, "Cryptocersus." In all these cases the protozoan gets shelter and food.

Mammals and Bacteria

In this association symbolic bacteria are present in the Caecum of rabbits and rumen of cattle. Here, the bacteria utilize the inorganic nitrogen and sulphates for the synthesis of essential amino acids, which can't be synthesized by the animal. In turn to mammals, the

Bacteria produce Vitamin B12 (in ruminants). In man also some of the vitamins are produced by the intestinal Bacteria. The Symbolic Bacteria are also present in the gut of the South African Honey Guide." Here, the Bacteria feeds on the Wax of honey comb the bacteria digest the Wax and gets shelter and food from the host.

Predation

In this relationship one animal is killed and eaten by another, the predator so predator is a free living animal. It is usually larger than prey. The herbivores that feed on the plant material completely are called as Plant Predators. There are, many kinds of predation which are not necessarily so selective. The swallow feeding on moths and midges, the top minnow or trout feeding on mosquitoes, and the ant eater feeding on an ant nest is less discriminate in taking individuals. Certainly, many able and healthy individuals are preyed upon. In cases such as these, which involve large numbers, the prey population often shows great productivity which can accommodate this mass mortality.

Many fish, insects, and other Invertebrates have tremendous production of eggs and larvae, but only a small percentage of these serve the onslaughts of predation and other forms of loss. For example, the horse shoe crab or Limulus of estuarine and evastal waters deposits its eggs on sandy beaches wehere they are consumed in great numbers by flocks of laughing gulls and other shore-birds. Large number of eggs also wash out to sea and are eaten by fish, whereas others are pushed up on the beach and dry out. Despite this fantastic loss of eggs and young larvae, Limulus continues to survive in large populations and has done so for the last 180 million years. It is one of the oldest animals on earth in terms of its evolutionary history, having outlived trilobites and dinosaurs by millions of years, and it is still incredibly abundant. There has been considerable interest in predation as a factor in quantitative population control as well as its role in quality control. On one hand, sportsmen have wanted to limit predation, being concerned that predators may take excessive numbers of game birds and wildlife. Some sportsmen's organizations have therefore supported predator control campaigns against hawks, owls, foxes, coyotes, wolves, cougar and other predators in the name of conservation. On the other hand, ecologists have often been interested in promoting predation as a method of biological control of pest animals, particularly insects and rodents, and even as a necessary biological control of desirable animals such as deer, or prevent them from becoming pests.

There are also examples from the entomological literature where predation serves as a significant population control on an animal population. In a classic study of natural population balance in the knapweed galifly (Urophora jaceana), a small dipteran insect, Varley concluded that one of the major controlling factors was the predation of mince upon insect pupae. His data showed that 22 to 43 percent of the fly morality was attributable to this predation. One study of tsetse flies (Glossing spp.) in Africa, found predatory spiders to be a controlling factor on the abundance and distribution of the flies. Insect predators have also been used in applied biological control of agricultural pests in a few cases. In California, the cottony cushion scale, an hemiptoran insect pest of citrus crops, was effectively controlled by a predatory ladybird bettle imported from Australia. In Hawaii, the sugarcane leaf hopper has been controlled by the capsid bug. Another important aspect of predation as an influence in population ecology and natural selection is the great variety of adaptations which have arisen in response to it. Protective coloration, warning coloration and mimicry fall into this category. Also, the morphological and behaviour, freezing behaviour and many other specialized patterns have developed in various species as adaptations to reduce mortality through predation. One particularly interesting system which has been studied in some detail is the intimate relationship between certain moths and the bats which prey upon them. The bats locate the moths by emitting ultrasonic pulses and detecting reflecting echoes flying moths, a natural sonar system.

The moths have evolved the ability to detect these ultrasonic pulses of the bats, and upon perceiving the approach of a feeding bat, they instantly go into complicatd diversionary flight spirals to avoid the bat. The introduction becomes on serial "dog flight" between bat and moth, with the moth trying to avoid the bat. This ability and behaviour has evolved in selective response to predation in the same way that cryptic coloration has evolved. In both case the advantageous mutations have been selected and propagated.

Parasitism

"Parasitism" is the relation between two individuals, wherein the parasite receives benefit at the expense of the host, therefore, parasitism is form of disoperation. In this association, the parasite receives (i) food (ii) shelter (iii) and protection, at the expense of the host. The host does not derive any benefit from the relationship.

Classification

Parasites are classified into different categories depending upon their relation with the host.

1. Temporary Parasites

The animals which spend only a part of their life cycle as parasites are called temporary parasites."For Example; The glochidium (is a Larva of freshwater mussel) leads a parasitic life. It attaches itself the skin of fish with the help of hooks present on the shell and later it is covered by the skin of fish. It remains burried thus for several weeks and emerges as young animals to lead a free life. Example 2. In prasitic Hymenoptera, the eggs are deposited into the body of another insect (Caterpillar), where the entire larvae stage is spent the larva feeds on the fat body of the host. Until it attains pupal stage. Finally it emerges as an adult and leads a free living life.

2. Permanent Parasites

Some parasites have continuous association with their hosts, they are called permanent parasites. They spend their entire life as parasites. For example: Trichinea worm, Malarial Parasite etc., Permanent prarasites often found among lower animals. They almost always require a succession of hosts or intermediat host.

3. Facultative Parasites

These are parasites (animals) which are parasitic only when there is a change for parasitic life, leading free life otherwise. Eg: Meggocs of some flies, Lamprey etc.,

4. Obligatory Parasite

Some species are forced to lead a Parasitic life. If not, they can't live. This type of parasites are called obligatory parasites. Eg: Filarial Wornm Cestodes etc.,

5. External Parasites

Those which live outside the body of the host are called external parasites. Eg. Flea on birds and mammals, blood sucking lice and fleas, mites, ticks and leeches.

6. Endoparasites

These are parasites which live in the elementary tract body cavities, various organs, blood or other tissue of the host. The association between 'endoparasites' and their hosts is invariably more inntimate than that between extoparasites and hosts. Tricchomonad flagellates, Opalnid ciliates Sporozoans, Nematodes, Cestodes and some Copepods are some of the endoparasitic forms. The endo-parasites are further divided into 2 types based upon their exact location inside the most body.

(i) Intracellular Parasites Ex:- Plasmodium (living within the cell)

(ii) Intercellular Parasites Ex:- Tryponasome (living be tween the cells in the intercellualar spaces).

Evolution of Paraitism

The origin of parasitic life in amimals is associated with attempts to secure food for themselves or their offspring. The ancestors of ecto-parasites were clearly free living forms. A free living organism could accidentally have settled on the body of a larger species where coditions were favourable for survival. For example: The biting lice might have evolved from the insects that live beneath the bark of trees. Endoparasites might have evolved from ectoparasites or directly from free-living ancestors.

Modifications and Adaptations

1. *Organs of attachment:* The rasping mouth of the Lampern, and the head of tape worm equipped with 'Suckers' and 'Hooks' for attachment to the hosts gutwall are the typical examples. Among parasitic arthropods such as mites other insects claws are well developed used to cling on the external surface of the host.

2. *Body Shape:* The intestinal parasites such as the tape worm and Round worm have elongated bodies. The liver fluke have leaf-like bodies and elongated form of blood flukes are good example of this type. These arrangements are useful to suit the environment in which the parasite lives.

3. Cuticles: *Ex: Tape worm.*

Intestinal parasites face the problem of with-standing the action of the digestive juices of the host. To avoid this these parasities develop thick cuticle, offers resistance against the action of the digestive juices.

4. Loss of Locomotary Organs:

Most of the endoparasites have lost their Locomotary organs, as they are not needed. There is no need in search of food, as they have always plentiful supply of food. So the locomotary organs of the parasites disappear or differentiate into organs of attachment. The fleas or lice (Parasitic Insects) have lost their wings, which live among the hair and feathers.

5. Digestive System:

In the tape worm, which lives in the alimentary canal of the hot, has lost its alimentary canal and absorbs already digested food from the surrounding medium. The mouth parts of the blood sucking insects have developed 'Piercing and Sucking' apparatus.

6. Loss of Sense Organs:

As the internal environment of the host is more or less uniform, there is no use of the sense organs among the endoparasites are provided only with tango receptors.

7. Complicated Reproduction:

Ex: Flat Worm, Round Worm, etc. The reproductive function in most parasites has undergone considerable complication. It is the goal of all the functions of the organism. The endoparasites are characterized by enormous fertility. For example (i) the human round worm lay (Ascaris Lumbricoides) eggs at a rate of 1500 a day (ii) the termatodes and cestodes lay eggs continuously (iii) Taenia solium (Tape Worm) lay eggs annually and have been estimated at 80 million (per year).

The most remarkable feature of flat Worm parasities is the presence of 'complicated reproduction'. The effect of "Parasitism" on the host is usually harmful. The endoprasites causes effects, more marked than those by extoparasites. The following Pathological effects caused by parasites in their hosts.

(1) The destruction of cells and tissues of the host by the parasite's movement or feeding act. Ex: Entamoeba histolytica-eats the tissue cells of colon and red blood cells of the host, and plasmodium feeds on liver cells and erythrocytes (R.B.C.)

(2) Parasites may cause 'enlargements's and disorders of lymph glands, and liver. Ex: Leishmania.

(3) Parasites may cause 'Malaria', (Plasmodium parasite) and 'Trypanosoma gambiense' diseases due to the secretion of toxins (Poisonous).

(4) The effects of endparasites may also be mechanical. Ex: Ascaris, is a good example when these parasites multiply in greater number they block the intestine, preventing the feree flow of food, which leads to the death of the host.

(5) The most remarkable effects of prasitism on the sexual characters are seen among crustaceans. The presence of sacculina bring spectacular changes in its host (Crab), which may result finally in death (due to the parasitic castration).

Hyperparasites:

Sometimes a parasite harbors a number of parasites in its body. They are called 'Hyperparasites'. A typical example is, small mammal ones as 'Mouse' which invariably harbours a no of ectoparasitic fleas, which themselves contain endoparasites (Protozoans) in their alimentary Canal.

Parthenogenic and Non Parthenogenic Parasites

The parasite which do not cause any injury to the host are called 'Non parthenogenic parasite'. The parasite which cause disease is known as "Parthenogenic Parasite". In man's digestive system 'Bacteria' are helpful in synthesizing' Vitamins, used to man. Some Bacteria causes numerous diseases as pneumonia, tuberculosis, diptheria, bubonic plague, cholera, syphilis and wooping cough etc. Viruses causes, small pox, polio, yellow fever etc. Animal-like microbes causes malaria, (Plasmodium) sleeeping sickness (Trypanosona gambiense) and dysentery (Entamoeba histolytica).

Parasitism is virtually universal in all plants and animals in vertebrate animals, endoparasites are found within many of the major organ systems of the body, most commonly in the digestive, circulatory, respiratory and urogenitial systems. Ectoparasites occur on or within the skin and its appendages, such as hair scales. Throughout most of the world, man is beset with many parasitic organisms including intestinal worms (tape-worms, flukes, roundworms, etc.), intestinal protozoa (amoeba, ciliates and flagelates), parasites of the blood (microfilaria, malaria protozoa, etc.), and a variety of extoparasites (Mice, mites, ticks, mosquitoes, etc.).

Many infectious microorganisms including bacteria and viruses are also parasites in the sense that they extract nutrients from the host. Some of these are pathogenic, that is, they impart the health and normal functioning of the host, as for example, the bacterial organisms of the genus shigella which cause bacillary dysentery. Others, such as some intestinal flagellates, are nonpathogenic and do not harm the host. Many parasitic organisms may be pathogenic in one individual and nonpathogenic in another. For example, most individuals, infected with Entamoeba histolytica in the digestive tract do not have amoebic dysentery, and, in fact, may have no disease symptoms at all, but in other individuals, the same organism may produce severe disease and even death. This is a matter of individual ecology and immunology which is not well understood. As mentioned previously in the selection on communalism, theere is no sharp line of distinction between obligatory commensalism and parasitism. Theoretically, Entamoeba histolytica feeding on tissues of the host is a parasite, whereas the flagellate Trichomores hominis, feeding on digested foods before incorporation into host tissue is not a parasite, nor is Entamoeba coli feeding on undigested particles of food and intestinal bacterias.

Neither is there a sharp line of distinction between a parasite and a predator. Houseflies and vampire bats bite and suck blood form livestock, and they might be considered either parasites or predators. Normally, we think of predators as killing their prey in a short time. Man's relation to the cow is predatory when he kills it for beef, but it is parasitic when he milks it for dairy products. There are some relationships, however, in which the distinction is arbitrary, as for example, in the lamprey feeding on a host fish. The lamprey may kill the fish in a few days or a few weeks, depending upon its size in relation to the fish, and may be considered either a predator or a prasite.

Ecologically, one of the most interesting questions about parasitism is the effect of the parasite population upon the host population. The oldest and best adapted parasites have little or no pathogenic effect upon the host. Pinworm infections in children are relatively harmless, though they may cause a minor irritation around the anus. In West Bengal, studies have shown that over 75 percent of the local population harbors hookworm (Necator sp. and Ancylostoma sp.), but there is no demonstrable pathology or effect on the people. The infectious burden (i.e., the number of parasites per person) seems to be small and insignificant. Most wild animals have parasites are able to maintian excellent health with these parasites. Zebras in Kenya are intensively infected with many internal and external parasites, but they are usually in remarkably robust health. On the other hand, there are numerous examples where parasitic infections are harmful and debiliting to the host. Certainly, hookworm necator americana in the southern United States, as it occurred 50 years ago, was a debilitating disease, sapping the strength and health of the infected person. Throughout the world, malaria still remains as a major health problem and it causes a great amount of illness and human misery.

One of the most important debilitating disease of modern times is schistosomiasis an infection of blood flukes of the genus Shistosoma. This is a waterborne infection and has spread throughout the tropical world with the expansion of irrigated agriculture. In animal populations, parasites may also weaken infected individuals. In the bighorn sheep of Wyoming and Idaho, lungworm infections are still a major cause of illness and mortality. In dogs and wolves, heartworms reduce the animal's vitality and hunting succss.

In meadow vole populations, botfly parasitism causes larger spleen weights in infected mice and reduces their survival when exposed to cold temperatures. Periodic reductions of field voles in Wales have attributed in some years to heavy infections with tuberculosis among the mice, but in other years similar reductions occurred without the infection. Similarly, reductions in populations of red grouse in Scotland have been attributed in some years to the round worm parasite, Trichastrongylosus pergracilis, but this has not a consistent finding in all year.

Occasional heavy mortality of gray squirrels in Baltimore has been associated with coccidiosis, caused by a protozoan parasite of the digestive tract. In 1971 a virus disease of wildfowl decimated 90 percent of the pheasants of England, in much the same way as myxomatosis reduced rabbit populations of England, and Europe in the 1950's. Thus, parasitism as an interspecific population relationship may be balanced or unbalanced in regard to its effect upon the host.

A seriously unbalanced relationship, in which the host dies is, of course, selectively disadvantageous to the parasite as well. Another outstanding feature of parasitism is the interspecific complexity which was evolved in some species. Many parasites have primary and secondary hosts and, in some cases, tertiary hosts for various stages of the life cycle. The malaria organism Plasmodium falciparum, for example has parts of its life cycle, the sexual phase of reproduction, within the mosquito, and part, the sexual phase of reproduction within the human host. Schistosomes infect man or other vertebrate animals as the primary host for sexual reproduction. The Chinese liver fluke, Clonorchis sinensis, infects man as a primary host, aquatic snails as a secondary host, and fresh-water fish as tertiary hosts. Despite the complexity of this life cycle, it is a very successful and widespread parasite, infecting millions of people throughout the Orient.

Competition

The word "competition" denotes striving for the same thing. At the ecological level competition becomes important when two organisms strive for something that is not in adequate supply for both of them. Thus, plants compete for light and nutrients in a forest, and animals compete for food and shelter when the latter are relatively scarce in terms of the density of animals. If the population consists of only a few scattered individuals, competition will not be a factor of ecological importance. In the arctic, for example, plants may be so few and scattered due to the severe climate that no competition for light occurs. The result of competition is that both parties (that is, the competition) are hampered in some manner. At the population level this means that density or rate of population energy flow, will be reduced or held in check by the competitive action. There is another type occurs when two organisms interfere with one another while striving for something even though that something is not in short supply. For example, the organisms might secrete substances that interfere with each other, or they might even eat each other. Many ecologists prefer not to include such direct mutual inhibition under the heading of competition.

Both intraspecific and interspecfic competition can be very important in determining the kinds and numbers of organisms. Intra-specific competition is an important factor in those populations that tend to be selfregulated in the manner described above. An interesting behaviour pattern, which results in intraspecific competition for space and a rather effective control of population size, is known as territoriality; it is characteristic of many species of birds and some other higher animals. At the beginning of the breeding season the male of a territorial species of bird will "stake out" a definite area of its habitat (that is, the "territory") and defend it against other males of the same species, with the result that no other ale is allowed to enter the area.

Much of the loud bird song we hear in the spring is for the purpose of announcing to other males "ownership" of land, and not for the purpose of wooing the female, as is often supposed. A male that is successful in holding his land has a high probability of mating and nesting, while a male that is unable to establish such a territory will not breed. Once the pair is formed, the female also joins in the defense of the territory in many species. The defended territory also serves a positive purpose of insuring the complex business of caring for the nest and young will not be interrupted by the presence of other males and females. Finally, it should be mentioned that the actual defense of the territory does not usually involve much actual fighting or other severe stress.

Would be invaders respect the established bird; loud songs or threat displays usually are sufficient to cause the invader to withdraw. However, if one of the pair should be killed it is very often quickly replaced by a bird form a reservoir of individuals not established. Therefore, since the territorial behaviour pattern helps avoid both overcrowding and undercrowding it can be siad to be regulatory in the sense that such behaviour promotes a sigmoid growth pattern, or at least helps prevent an overshoot of carrying capacity.

Plant populations, as well as animal populations, may regulate themselves, so to speak, and avoid overcrowding. As one drives through the desert in the southwestern United States one is impressed with the fact that desert shrubs are widely spaced, often almost uniformly distributed as if planted at regularly spaced intervals. The pattern would seem to be explained by competition for scarce water, which eliminates all but one individual in a given area. However, evolution of self-regulation of the population has occurred in some species, in that severe competition for water is avoided by the production of leaf or root hormones that inhibit development of other individuals in the neighborhood. That is, chemical substances rleased by decaying leaves that fall to the ground under the shrub, or by the living root in the soil, inhibit or kill any seedlings that may start to sprout.

Such a control mechanism would tend to keep plants spaced apart, thereby reducing actual competition for water, which might result in stunting or death of all plants should there be many trying to grow in the same place. In this kind of "birth control" the quality of the individual in a limited environment is maximized, not the quantity; is there as lesson here for man?

Amensalism and Antibiosis

Amensalism is an interspecific relationship in which one population is inhibited while the other is unaffected. A simple example is the shading out of certain plants under tall trees. The trees reduce the available sunshine at the ground level, and numerous species of plants cannot find adequate light in the shade. Hence, only shade-tolerant plants with lower light requirements can survive as ground cover in the forest. Antibiosis is a specific type of amensalism in which one organism produces a metabolite that is toxic to other organisms. The best known example is the mold, Penicillium, which produces an antibiotic substances causing the death of many bacteria. This was observed and identified in the classic work of Florey and Fleming in England in the late 1920's and early 1980's which led to the development of antibiotics in clinical medicine. Penicillin, strepotomycin, aureomycin and other antibiotics used against disease organisms all represent a similar ecologic principle, and one which is widespread in nature.

Many species of fungi and lichens produce metabolic substances which inhibit bacterial growth. The green algae Chlorella produces substances which inhibit the growth of diatoms of the genus Nitzschia, and conversely, Nitzschia produces a substance which inhibits Chlorella. Many plants also produce substances which are toxic or inhibitory to animals. Old cells of Chlorella produce a substance which inhibits the feeding of Daphnia. Algal blooms of some blue-green and red algae produce chemicals toxic to fish. The well known Red tide of southern coastal waters, produced by the flagellate Gymnodinium brevis, can result in massive fish kills. Land plants also produce inhibitory substances. The roots of some species of trees produce substances which inhibits the growth of other trees, both of their own species and different species. This may result in a characteristic spacing pattern of trees within a forest, dependent not so much upon competition for sunlight at the conopy level, but upon competition for space, water or nutrients at the subterranean root level. Many plants produce substances toxic to animals which touch them or eat them.

In Nevada and Utah the desert bush Halogeton glameratus produces oxalic acid which is poisonous to sheep. Poisonous toadstools to and tubers have been well known throughout the history of man. In South America and Africa, many native people harvest the tubers of the poisonous manioc plant, but they place it in a series of soaking vats to leach out the toxic chemicals before consuming the starchy tissue of the roots. A small African monkey, the talapoin monkey (*Cercopithecus talapoin*), has learned to steal the manioc from the soaking vats at just the proper state. Many of these substances are in a class of compounds known as ectorines of environmental harmones. Ectorines are substances produced by the organism which affect other organisms of the same or different species. Not all ectocrines are toxic or inhibitory, of course; many are stimulatory or beneficial.

Aggregations

All organisms react on their habitat in one way or another, and when they occur in numbers these reactions produce a conspicuous effect. Water conditioning occurs hence physical or chemical changes occur as the result of organisms living in it. Compared with unconditioned water, these changes may have either a harmful or a beneficial effect on organisms introduced into the water after the original organisms have been removed. Water is said to be homotypically conditioned when the changes were previously produced by individuals of the same species as being studied and heterotypically conditioned when the changes were produced by a different species. Experimental studies have demonstrated that gold-fish grow faster in water that has been homotypicaly conditioned for 24 hours than in unconditioned water. Both fish and amphibian larvae also do better in water conditioned by the presence of mollusknes than in unconditioned water.

The marine flatworm Procerodes wheatland will survive much longer when transferred to fresh water conditioned by the presence of either live or dead individuals of the same species or by freshwater species of flatworms than they do in unconditioned fresh water. The longer survival in toxic solutions, faster growth, and greater reproduction of protozoans, snails, flatworms, cladcerans, amphibian larvae, and fish occurring in aggregations rather than as isolated individuals is attributable to water conditioning. Benefits derived from aggregating are shown in other ways by terrestrial animals. In honeybees, when hive temperatures drop below 14°C (57°F) during the winter, they form clusters and maintain a mass temperature several degrees above outside temperatures. This is brought about by increased metabolic oxidation of honey in their bodies and by increased muscular activity.

Furthermore, the compact cluster presents a surface area for heat loss that is less than the total surface area of the individuals separately. When there is danger of overheating, the bees in the hive spread out on the combs and fan with their wings to create a circulation of air. They will also carry water into the hive and place small quantities both outside and inside the comb cells. The forced air circulation evaporates the water and cools the hive. Bees also cool themselves by constantly moving their tongues in and out of their mouths, exposing to evaporation the moisture that is present on them as a thin film temperature regulation is less well developed in other social Hymenoptera.

Allelochemistry

Included here are the coactions whereby chemicals secreted by one organism affects the growth, health, or behaviour of other organisms. Allelopathy is produced in plants when toxins are liberated that inhibit seedling growth in the vicinity. This may affect succession of plant species, especially important in the early stages. Some pioneer species in the abandoned field sere produce substances inhibitory to nitorgen-fixing and nitrifying bacteria. This retards invasion of other species that require higher nitrogen concentration in the soil.

Volatile inhibitors are generally more prevalent than water-soluble ones, and relatively more prevalent in arid than humid climates. Antibiotics produced by bacteria, fungi, actinomycetes, and lichens are widespread in nature and may be one of the reasons why bacteria pathogenic to man cannot mutiply well in soils. A number of antibiotics, such as penicillin, have been used extensively in human medicine. The use of allelochemic coactions agriculture has possibilities but has yet to be exploited. Among animals, overcrowding of tadpoles in culture dishes in associated with the occurrence of peculiar round vacuolated cells in the intestinal tract and feces that appear responsible for curtailment of further growth. Under laboratory conditions, killer stocks of Paramecium aurelia produce a toxin, paramecin, at the rate of one unit-particle per animals per 5 hours. One unit-particle is enough to kill one individual of sensitive stock of the same species as well as being lethal to other species of Paramecium. Conditioning that becomes unfavourable homo-typically may sometimes be favorable, or at least tolerable, heterotypically. Thus in protozoan infusions there is a microsere of one species succeeding another. Allechemic effects are of great variety in both plants and animals: repellants, escape substances, suppressants, venoms, inductants, attractants, signals, stimulants, autotoxins, antoinhibitors, and so on. Pheromones chemical messages between members of a species especially important in reproductive behaviour, social regulation and recognition, alarm and defense, territory and trail marking, food location, as so on Many of these effects are beneficial to the individual; others serve for competitive purposes.

Protocooperation

In the positive relationship between populations, proto-cooperations is a short step ahead commensalism, to cooperation, both populations gain by the association or interaction of some kind. A considerable work on protocooperation was one by Alee. In nature, crabs and coelenterates (sea anemone) are often found associated with each other. This association is of mutual benefit to both. The coelenterates grow on the back of the crabs, or sometimes planted there by the crabs. The coelenterates provide protection to the crabs. In turn, these are transported by crabs to new situations, which ensure them good supply of food. There is no physiological relationship betweeen the coelenterate and the crab, and none is absolutely dependent upon the other.There exists complex relationships between different populations. Some of these relashionships play a vital role in the regulation of populations.

3 Fluctuations

Populations of similar species in the same place may respond to different environmental factors. For example, the densities of four species of moths, whose larvae feed upon pine needles, fluctuated more or less independently over a 60-year period. The populations were sampled by counting the number of pupae (or hibernating larvae in Dendrolimus) per square meter of forest· floor in a managed pine forest in Germany. The populations varied over three to five orders of magnitude with irregular periods of a few years.

Furthermore, the highs and loss of the populations did not coincide closely, suggesting that even though the species fed on the same resource in the same forest, their populations were governed independently by different factors. Weather and food supply seem unlikely agents, as these should have affected each species similarly population regulation may possibly have involved specialized predators or parasites. Some populations of birds and mammals undergo periodic cycles of abundance. Population cycles have contributed to the lore of population ecology since Charles Elton's paper "Periodic Fluctuations in the Numbers of Animals: Their Causes and Effects," published in the *British Journal of Experimental Biology* in 1924. Most of Elton's data concerned fur-bearing mammals in the Canadian boreal forests and tundra, where the Hudson's Bay Company had kept detailed records of the numbers of furs brought in by trappers each year. Data for the lynx and its principal prey, the snowshoe hare, revealed regular fluctuations of great magnitude (Figure 18-4). Each cycle lasts approximately 10 years and cycles of the two species are highly synchronized, with peaks in lynx abundance tending to trail those in hare abundance by a year or two. The periods of population cycles vary from species to species, and even within a species.

In Canada, most cycles have periods of either 9 to 10 years or of 4 years. Although colored foxes have a ten-year cycle over most of their range, they exhibit a pronounced 4-year cycle in Labrador and on the Ungava Peninsula, where they prey mostly on lemmings, which have 4-year cycles as well. In general, small herbivores, such as voles and lemmings, have 4-year cycles; large herbivores—snowshoe hares, muskrat, ruffed grouse, and ptarmigan—have 9- to 10-year cycles. Predators that feed on short-period herbivores (Arctic fox, rough-legged hawk, snowy owl) themselves have short population cycles. Predators of larger herbivores (red fox, lynx, marten, mink, goshawk, horned owl) have longer cycles. The length of the cycle also appears related to habitat, with longer periods in forest-dwelling species and shorter ones in tundra-dwelling species. Populations may cycle quite regularly. For example, a string of seven 4-year periods in the colored for returns to the Moravian missions between 1847 and 1880 prompted Elton to remark that an Ekimo hunter "might have reflected that his good luck had his bad luck chased each other with sufficient regularity to amount to a natural law."

But frequent variations in the periods between peaks as well as irregularities in peak and low numbers convinced some observers that population fluctuations reflect random processes, not regular cycles. Lamont Cole dispelled this notion by showing that the distribution of periods between population peaks differed significantly from that expected for series of random numbers. To derive the random distribution, Cole reasoned that turning points in a series of random increments-that is, either a high point or a low point-occur with a probability of one-half for each time interval. For example, an increase in one interval is followed by another increase with probability 0.5, and by a decrease (defining a peak between the two) with probability 0.5. Furthermore, one-half the turning points are maxima.

Using probability theory, Cole calculated the expected proportions of periods between peaks of two, three, four,... intervals. The expected mean interval of the random series is exactly 3 years. The average interval between peaks in fluctuating mammal populations is characteristically greater than 3 and the distributions of intervals differ markedly from the expectation for a random series. Therefore, population cycles cannot be dismissed simply as the result of random processes. In the wake of the controversy created by Andrewartha and Birch, most ecologists realised that populations vary in response to both density-de-pendent and density-independent factors. Subsequent research focused on identification of the most important factors and on learning how their effects were expressed through population processes.

Even before Smith' critique of Andrewarth's analysis of the thrip population, Canadian and English entomologists concerned with the control of insect pests and begun to apply new methods of analysis to identify factors that caused populations of insects to vary. In the introduction to his paper "Single-factor Analysis in Population Dynamics," R. F. Morris, of the New Brunswick Forest Biology Laboratory, explained his purpose:

> A preliminary examination of rather extensive life-table data for the spruce budworm... suggested that the factors affecting this species in any one place are of two types - those that cause a relatively constant mortality from year to year and contribute little to population variation, and those that cause a variable, though perhaps much smaller, mortality and appear to be largely responsible for the observed changes in population.... A factor of the latter type will here be called a "key factor," meaning simply that changes in population density from generation to generation are closely related to the degree of mortality caused by this factor, which therefore has predictive value.

Morris conceived of changes in insect populations by a simple model,

$$N(t + 1) = N(t)FS_1S_2S_3$$

in which the factor of population increase from year t to t + 1f is the product of fecundity (F) and survival with respect to potential mortality factors (S). In logarithmic form,

$$\log N(t + 1) = \log N(t) + \log F + \log S_1 + \log S_2 + ...$$

He wished to determine if variation in $\log N(t+1)$ followed upon variation in one or a few of the values of F and Si. He reasoned that variation in one or more of these variables reduced the correlation between $N(t + 1)$ and $N(t)$. But if S_2, for example, were responsible for changes in N from year to year, $N(t + 1)$ would be more strongly correlated with $N(t)S_2$ than with $N(t)$ alone. Morris applied his analysis to data for the black-headed budworm.

The budworm belongs to the moth family Tortricidae (leaf-rollers), native to eastern Canada, and is a major defoliator of fir trees in New Brunswick. One generation of adults appears each year; the population overwinters in the egg stage. Twelve years of observations on the density of budworm populations and the percentage of parasitism of the larvae by wasps and flies revealed a fluctuation in numbers of over two orders of magnitude with a period of about 9 years.

The survival of larvae at risk to parasites (S) followed a similar course, being high during the population buildup and low during periods of decline. In this example, the correlation (r) between $N(t + 1)$ and $N(t)$ was 0.67, indicating that 45 per cent (r^2) of the variation in population size was related to variation in population size during the previous year. When $N(t)$ was multiplied by S. the correlation between $N(t + 1)$ and $N(t)S$ was 0.93, and the proportion of variation explained was 86 per cent. From this analysis, Morris concluded that larval parasitism was a key factor in the population processes of the black-headed budworm.

English entomologists G.C. Varley and G.R. Gradwell extended Morris's analysis to examine the effects of many factors and determine statistically which contributed to variation in population size. Following the terminology of J. B. S. Haldane, Varley and Gradwell defined the "killing power" of a particular agent of mortality i as k_i = - $\log S_i$, and the total killing power exerted upon a population as the sum of each component, $K = k_1 + k_2 + k_3 +...$ (The expression log would have served equally well.) Their approach then followed directly from Morris's in using correlation and regression to determine which variables were most strongly correlated with variation in [$\log N(t + 1) - \log N(t)$ or, equivalently, K. Because of their particular terminology, Varley and Gradwell's approach is often referred to as "k-factor analysis." During the 1960s, entomologists disputed the relative merits of the Morris and Varley-Gradwell approaches to key-factor analysis. finally resolving the issue by the end of the decade, by which time Morris's analysis was less persuasive.

The dispute is unimportant now, except historically, because it arose from unknowledgeable application of statistics and not from any disagreement over biological issues. Regardless of such statistical subtleties, the application of Morris's and Varley-Gradwell's analysis to insect populations has led to identification of stages of the life history during which demographic processes contribute most to variation in population size. The analysis by Canadian economic entomologist D. G. Harcourt of the diamondback moth, a worldwide pest of cabbage and related crops, is typical of such studies.

In southern Ontario, the diamondback moth usually has four, but exceptionally up to six, generations per year. During each growing season between 1958 and 1961, Harcourt sampled populations at each stage of the life cycle (egg, several larval instars, pupa, and adult). He determined the rate of parasitism from extensive rearings of larve and pupae, and estimated fecundity by raising adult months in the laboratory; mortality of the adults was adults from the previous generation had survived to lay eggs.

Mortality during each stage varied from as little as 1 per cent of eggs to 89 per cent of adults (that is, 89 per cent of the potential fecundity of adult females). The major causes of mortality also varied from one stage of the life history to the next. Small larvae were particularly vulnerable to heavy rainfall, whereas large larvae and pupae were heavily parasitized by wasps. Adults usually succumbed to weather factors. Each of the causes of mortality itself varied considerably from generation to generation. Harcourt assessed the contribution of variation in survival rate during each stage to variation in numbers in the population at the beginning of each generation by correlating [log$N(t + 1)$ - log($N(t)$] with each of the survival rates.

Not surprisingly, 73 per cent of the variation in *N* derived from variation in adult survival; mortality rate was greatest during that stage and its variation more or less governed the population trend. But high mortality does not necessarily imply cause of variation in *N* between generations. Rainfall during the first larval period caused the death of 55 per cent of the individuals in the population, but variation in the factor was not significantly related to the trend in the population from generation to generation. In these diverse insects, the critical stage at which key factors acted, as well as the nature of the key factor itself, varied from species to species. In only one of the species, the Colorado potato beetle, was food supply a key factor. Weather was the most important factor in three of the species, including the diamondback moth; and disease, parasites, and emigration from local populations ranked highly in the rest. In 9 of the 12 studies, the key factor was determined to be density-dependent in the sense that its value was found to vary inversely with the initial size of the population.

The fluctuations in populations of insects, and other organisms, may be thought of as resulting from changes in birth and death rates brought about by changes in the environment. When a population is regulated by a strong density-dependent factor during at least one stage of the life history, even a density-independent change in fecundity or survival will bring density-re-strong mechanisms into play and reveal the density-dependent factor. Changes in density-dependent factors alter the value of *K* itself and thereby put the population out of equilibrium. In either case, the population is continually in the process of returning to *K* or catching up with a varying value of *K*. How well it does so depends on the capacity of the population to grow relative to the period of variation in density-independent factors or in *K*.

Populations with high potential growth rates achieve their equilibria quickly, and thus track variation in *K* closely. Populations with low growth potential cannot take advantage of increases in *K* and are also likely to resist decreases in *K*. Hence such populations vary little.

Population biologists compare the response of populations to change by their characteristic return times (*T*). *T* is simply the inverse of the exponential rate of growth of a population free of the effects of crowding; hence $T = 1/r$. Theoretical analyses have shown that a population will track the environment closely when its characteristic return time is less than the period of the environmental fluctuation (time from peak to peak) divided by 2 (Nisbet and Gurney 1982). When *T* is much longer than the period of the pertinent environmental variation, the population varies little. Small insects have values of *T* much shorter than 1 year and their populations track seasonal variation in conditions; the same variation evokes little response from populations of long-lived mammals.

Except for factors associated with daily, lunar (tidal), and seasonal cycles, environmental fluctuations tend to be irregular rather than periodic. Historical records reveal that years of abundant, extreme heat or cold, or such natural disasters as fires and hurricanes occur irregularly, perhaps even at random. Biological responses to these factors are similarly a periodic. For example, widths of the growth rings of threes vary in direct relation to temperature and rainfall; patterns of successive ring widths cannot be distinguished from a random series. Trends in the sizes of many populations do, however, change with periodic frequency. For many years, ecologists believed that explanations for such cycles must be sought among environmental factors exhibiting similar periodic variation.

The regular 11-year cycle in sunspot numbers was frequently mentioned, but the sunspot cycle never matched population cycles well and no model could be devised to link the two. By 1940, the idea had been abandoned. With the development of population models in the 1920s and 1930s it became evident that because of their inherent dynamical properties, populations subjected to even minor, random environmental fluctuation could be caused to oscillate.

Such cycling can result from time delays in the response of births and deaths to changes in the environment. Just as the momentum imparted to a pendulum by the acceleration of gravity carries it past the equilibrium point and causes it to swing back and forth periodically, the "momentum" imparted to a population by high birth rates at low density or high death rates at high density carries the population past it equilibrium when demographic responses are time-delayed.

Cyclic population behaviour has been explored in a number of theoretical treatments, largely confirmed by laboratory experiments. Time delays that cause populations to oscillate when displaced from their equilibria are inherent to models based upon discrete generations. According to these models, populations respond by discrete increments from one time to the next and therefore cannot continuously readjust growth rate as population size approaches equilibrium. This can cause the population into overshoot the equilibrium, first in one direction and then the other as N draws closer to K. In models of continuously growing populations, oscillations can be induced by introducing a time lag in the term for density dependence.

Populations with discrete generations may be described as growing according to the expression $N(t + 1) = N(t)\lambda$ or, in logarithmic from $\text{Log}N(t + 1) = \text{Log}N(t) +$

log. According to logistic growth, logλ decreases in direct relation to the size of the population; that is, logλ = r[1 - $N(t)/K$]decreases in direct relation to the size of the population; that is, log, where r is the intrinsic exponential growth rate of the population free of density effects. Now we may write an equation for discrete logistic growth as

$$\log N(t + 1) - \log N(t) = r(1 - N(t)/K)$$

The behaviour of a population growing accordingly is difficult to characterize because equation has terms with both log$N$ and N; that is, it is not linear in N. In such circumstances, one may approximate the equation in the region close to the equilibrium point by a linear form. In the case of above equation, as $N(t)$ nears K, [1-$N(t)/K$] approaches [log$K$ - log$N(t)$]. Therefore, close to the equilibrium point K, change in population size may be described reasonably well by

$$\log N(t + 1) -- \log N(t) = r[\log K -- \log N(t)]$$

This equation reveals that the behaviour of the population near the equilibrium depends critically on the value of r. When r is less than 1, the increase in the population between t and t + 1 will be less than the difference between the population size and the equilibrium. Therefore, the population will approach the equilibrium directly. When r exceeds 1 but is less than 2, the population will overshoot the equilibrium, because log$N$ exceeds the difference between log$K$ and log$N$, but it will end up closer to the equilibrium than before. Thus the population will oscillate back and forth across the equilibrium value, getting closer with each generation. This behaviour is called damped oscillation. When r exceeds 2, the population ends up farther from the equilibrium each generation and the oscillations increase. With increasing r, these oscillations take on very complex, eventually unpredictable forms referred to as "chaos," The full range of behavior of this model and a variety of related ones is described by May, Man and Oster, Schaffer, Pool, and, in a very thorough presentation, by Nisbet and Gurney. Briefly, as deviations from the equilibrium become large, we can no longer deal with the linearized form of equation and must encompass its full expression in our thinking. This step will lead us to the concept of the limit cycle.

The discrete logistic equation,

$$N(t + 1) = N(t)e^{r[1 - N(t)/K]}$$

is portrayed as plots of $N(t + 1)$ versus $N(t)$, and $N(t)$ versus t, for values of r of 0.5, 1.5, and 2.5, in Figure 18-9. When is less than 1 (top graphs), for values of $N(t)$ less than K, values of $N(t +1)$ are also less than K, and the population increases (or decreases) monotonically toward the equilibrium. When the value of r is between 1 and 2, for some values of $N(t)$ less than K, $N(t + 1)$ exceeds K and the population overshoots. The oscillation is nonetheless damped and the population eventually reaches the equilibrium, as we determined from our analysis of the behavior of the model close to the equilibrium. The position of the maximum of $N(t+1)$ with respect to $N(t)$, determined by differentiating equation (18-3), is $N(t) = K/r$, at which point $N(t + 1) = (K/r)e^{r-1}$. Inspection of the graph shows that oscillations will be damped so long as the slope of the relationship between $N(t +1)$ and $N(t)$ is less negative than - 1 at K.

Under this condition, regardless of the overshoot, the population at $t + 1$ will always be closer to K than it was at t. The slope $dN(t + 1)/dN(t)$ at K is simply (1 - r); thus it is less negative than -1 when r is less than 2. When r exceeds 2, however, $dN(t + 1)/dN(t)$ is more negative than - 1 and the population moves farther from the equilibrium - up to a point. Below K, the curve relating $N(t + 1)$ to $N(t)$ eventually bends back down toward the origin of the graph and, above K, the curve eventually flattens out as $N(t + 1)$ approaches 0. As a result, a line with slope - 1 passing through $N(t + 1) = N(t) = K$ crosses the curve at two points corresponding, let us say, to N and N.

Clockwise of this line, the population exhibits increasing oscillations, but counterclockwise of the line the oscillations are damped. This means that close to the equilibrium the oscillations increase, but at high amplitude the oscillations are damped. The result is a stable oscillation, with the population eventually alternating between N and N. The models described above pertain to simple populations without age structure. In the more complicated case, fluctuations in a population are accompanied by changes in the age distribution of individuals and, as a result, the projection of population size becomes very difficult. P. H. Leslie examined the behavior of an age-structured population with density-dependence using a projection matrix. In his particular example with four age classes ($x = 0, 1, 2, 3$), the population exhibited a damped oscillation with a period of about 9 to 10 years.

Oscillations are produced in continuous-time models when the response of population growth to density is time delayed; that is, when the effect of density dependence reflects the density of the population t time units i the past (t is the lower-case Greek letter tau). Modified thusly, the logistic equation becomes G. E. Hutchinson pointed out that this model produces damped oscillations in N so long as the product rt is less than Below the population increases or decreased monotonically-without oscillation-to the equilibrium point. For rt greater than the oscillations increase until the maximum population size reaches $N/K = e$. Thus for $rt = 2$, oscillations increase in amplitude until the maximum value of N is times K.

Population biologists refer to such stably maintained oscillations as limit cycles. Their periods, from peak to peak, increase from about 4 times t to more than 5 times t with increasing r. Population cycles have been observed in many laboratory cultures of single species. David Pratt's observations on the water flea Daphnia magna have been widely quoted, partly because the populations exhibited marked oscillations when cultured at 25°C but strong damping at 18°C. The period at 25°C appeared to be just over 40 days for two cycles, suggesting a time delay in the density-dependent response of about 10 days; this is on the order of the average age of water fleas giving birth at 25°C. The time lag arose in the following manner.

As population density increased, reproduction decreased, to near zero when the population exceeded 50 individuals. In contrast to fecundity, survival was less sensitive to density and adults lived at least 10 days, even at the highest densities.

As a result, the pulse of deaths in the population lagged about 10 days behind the pulse of births at the beginning of each upswing of the cycle With births prevented by high density early in the cycle, when the population fell to densities low enough to permit reproduction it contained only senescent, nonreproducing individuals. Thus the beginning of a new cycle awaited the buildup of young, fecund individuals. The length of the time delay was approximately the average adult life span at high density.

At the lower temperature, reproductive rate fell quickly with increasing density, and life span increased greatly over that at 25°C at all densities. Populations at the colder temperature apparently lacked at a time delay because death was more evenly distributed over ages; and some individuals gave birth, even at high population densities; consequently generations overlapped more broadly. At the higher temperature, the Daphnia behaved according to a discrete-generation model with its built-in time delay of one generation. At lower temperature, they behaved according to a continuous generation model with little or no time delay.

Clyde Goulden and his colleagues at the Academy of Natural Sciences of Philadelphia have shown that the storage of lipid reserves by some species of water fleas reduces the sensitivity of mortality to density and therefore introduces a time delay into the population processes. *Daphnia galeata,* large species, stores energy in the form of lipid droplets during periods of abundant food (that is, low density), which it can then live off of when food supplies are reduced by overgrazing at high population density.

Females also pass lipid to each offspring through oil droplets in the eggs, thereby increasing the survival of young, prereproductive water fleas under poor feeding conditions. The smaller Bosmina longirostris stores little lipid and therefore starvation increases directly in response to increases in population density. The consequences for population growth are predictable: Daphnia exhibits pronounced limit cycles with a period of 15 to 20 days under the conditions of the experiment; Bosmina populations grow quickly to an equilibrium with perhaps a single strongly damped overshoot.

For Damphnia, goulden *et al.* estimated that r was about 0.3 days. With a cycle period of 15 to 20 days, t must have been about 4 to 5 days, and therefore rt was about 1.2 to 1.5. Because the value of rt was somewhat less than $\pi/2$, the cycles in the Daaphia population should have damped out eventually.

The behaviour of a population with respect to its equilibrium is sensitive to many aspects of life history that govern time delays in responses to density Slight differences in culture conditions or the intrinsic properties of species can tip the balance between a monotonic approach to equilibrium and a limit cycle.

A. J. Nicholson's experimental manipulation of time delay in laboratory cultures of the sheep blowfly Lucilia cuprina provide a dramatic demonstration of the relationship of time delays to population cycles. Under one set of culture conditions, Nicholson provided the larvae with 50 grams of live per day while giving the adults unlimited food. The number of adults in the population cycled through a maximum of about 4000 to a minimum of 0 (at which point all the individuals were either eggs or larvae), with a period of between 30 and 40 days. In this experiment, regular fluctuations of the blowly populations were caused by a time delay in the response of fecundity and mortality to the density of adults in the cages.

When adults were numerous, many eggs were laid, resulting in strong larval competition on for the limited food supply. None of the larvae that hatched from eggs laid during adult population peaks survived, primarily because they did not grow large enough to pupate. Therefore, large adult populations gave rise to few adult progeny, and because adults lived less than 4 weeks the population soon began to decline.

Eventually so few eggs were laid on any particular day that most of the larvae survived, and the size of the adult population began to increase again. Nicholson's result may be interpreted as a time-delayed logistic process, which provides a good fit to the observed oscillations in one of Nicholson's experiments with a value of rt = 2.1. This model predicts the ratio of the maximum to the minimum population to be 84 and the cycle period to be 4.54t. Nisbet and Gurney devote an entire chapter to the analysis of Nicholson's data, ending up with a somewhat more complex model to account for some of the departures of the cycles from the time-delayed logistic model. R. W. Poole also presents a detailed treatment of the data based on forecasting from time-series data. Regardless of these details, however, the experiment clearly reveals that density-dependent factors did not immediately affect the mortality rates of adults as the population increased, but were felt a week or so later when the progency were larvae.

Larval mortality was not expressed in the size of the adult population until those larvae emerged as adults about 2 weeks after eggs were laid. The blowfly population resembled Pratt's Daphnia population at high temperature in which crowding created discrete, nonoverlaping generations of individuals with an inherent time delay equal to the larval development period, about 10 days. Michael Hassell, John H. Lawton, and Robert M. May explored discrete generation models of the blowfly population; their paper should be consulted for further discussion. The hypothesis that population cycles were caused by time delays could be tested directly by eliminating the time delay in the density-dependent response; that is, by making the deleterious effects of resource depletion at high density felt immediately. Nicholson did this by adjusting the amount of food so that food limited adults as severely as larvae. Because adults require protein to produce eggs, by restricting the liver available to adults to 1 gram per day, Nicholson cut egg production to a level determined by the availability of liver rather than the number of adults in the population.

Under these conditions, the recruitment of new individuals into the population was determined at the egg-laying stage by the influence of food supply on per capita fecundity, and most of the larvae survived. As a result, fluctuations in the population all but disappeared. We have seen that responses of populations to density can be delayed by development time and by the storage of nutrients, both of which put off deaths to a later point in the life cycle or to a later time. Density-dependent effects on fecundity can act with little delay when eggs are produced quickly from resources accumulated over a short period. Populations controlled primarily by such factors would not be expected to exhibit marked oscillations.

Time delays longer than a single generation may be introduced through the transmission of maternal effects to progeny through the egg or by selection of different genotypes at different population densities. Studies on the western tent caterpillar by W. G. Wellington suggested a time delay through maternal effects. A 4-year survey on Vancouver Island, Canada, included a peak year for the population, 1956, followed by a rapid decline through 1959.

The larvae of the tent caterpillar were classified as either "active" or, "sluggish" depending upon their behavior, which, in turn, was determined by the level of nutrition of their mother. Broods of tent caterpillars composed primarily of active larvae differed from those with a higher proportion of sluggish larvae in that they constructed more tents with a more elongate structure, foraged over greater distances along the limbs of the tree, the more leaves, and as a result of their greater food intake, developed more rapidly. Active caterpillars survived better than sluggish ones. The proportion of active larvae in a particular brood was influenced by the past history of the infestation. Broods in areas in which the population had been low in previous years were likely to contain more active caterpillars than broods from areas of recent high levels of infestation.

The activity level of the larvae depended on the nutrition provided by the eggs from which they developed. Most moths laid similar numbers of eggs, regardless of their ability to provision each individual egg; how well each female provisioned her eggs with nutrients depended on how much she consumed as a caterpillar. When food was relatively scarce, few eggs were properly provisioned and broods contained few active caterpillars.

Thus the quality of caterpillars in broods of a particular year depended on the availability of food to the female parent when she was a caterpillar during the previous year. Furthermore, because the activity level of each caterpillar partly determined how much food it consumed, the food supply of its parent during the previous year affected the success of its progeny during the following year. In this way, populations could exhibit a time delay of more than 1 year in their response to density or to changes in the availability of food. Regardless of the time delay in the density-dependent response, a population at its equilibrium point will remain there until perturbed by some outside influence, whether a change in the equilibrium level (K) or a catastrophic change in population size (N). Once displaced from the equilibrium, some populations will move toward stable limit cycles, depending on the nature of the time delay and the response time. Others will return to the equilibrium directly or though damped oscillations.

4 POPULATION REGULATION

Many of the problems experienced in accommo-dating a growing family in a one-room house on a small plot ae analogous to those posed by expanding population. The size of the earth is fixed and its resources are strictly finite. Demands upon natural resources expand rapidly while their per capita availability decreases progressively, damaging the natural resource base. The consequences of environmental degradation and resource scarcity hit the poor most severely. To stay alive these people destroy the very resources they will need tomorrow. The resultant damage to the environment only depend their poverty. The vivious circle of poverty and environmental destruction is at work. The 'scissors effect' of poverty and increasing population slices away their ability to sustain human life.

The victims of poverty destroy forests for fuel-wood-, food, water and fodder and the growing population forces them to farm marginal land at non-sustainable levels. The excessive resource exploitation combined with a poor understanding of the interrelationships between man and environment has thus led to ecological crisis. Soil erosion and land degradation, deforestation, the fuelwood shortage, problems of water management, excessive population pressure on land and other environment-degrading trends inhibit the pace of economic development. The mounting problems of pollution consequent upon unimaginative interference of human beings with nature have been adversely affecting the quality of life. Pollutants may be regarded as resources at the wrong place, but as such they represent a continuous drain on resources, apart from their disruptive or degrading impact on air, soil and water. However, the harsh reality of the environmental crisis-shinking of forests, increasing desertification, degradation of critical resources, loss of agricultural land, salanisation, alkalisation and waterlogging, acidification of the environment and release of hazardous waste has created environmental awareness among planners and policy makers.

Building environmental concerns into development is now regarded as making efforts for sustained life and for avoiding the imminent death trap. The concept of sustainability has emerged as a key issue in development-planning very recently. The Stockholm Conference on the Human Environment 1972 and the International Conference on Population held in 1948 at Mexico City reflected concern for sustainable development.

The International planned Parenthood Federation and the international Union for Conservation of Nature ad Natural Resources have also stressed this approach to development. The recent report or the World Commission on Environment and Development, headed by the Norwegian Prime Minister Mrs. Brundtland has used it as a key concept. The need for sustainable development presupposes that the present development is not sustainable, and hence it is urged to opt for a new approach to development. There is need to integrate population, environment and natural resource policies into national development strategies so as to ensure environmentally sound sustainable development.

The major emphasis of this approach is on meeting the basic needs of the people and improving their quality of life without adversely affecting the viability of the environment. It requires development strategies which anticipate environmental problems and take precautionary measures to avoid the resultant problems. However, the goal of sustainable development is not likely to be attained unless the implications off interrelationships among population, environ-ment, resources and development are adequately understood and appreciated by the people of the country. It is urgently required to removed the insularity of perception of growing degradation of environment and critical resources in the context of rapid population growth and the process of development. This is possible when a proper understanding of environmental problems and their possible solutions is developed.

An attempt, therefore, is being made in the following pages to discus major environmental problems such as deforestation, soil erosion, flooding, excessive ground water exploitation, threat to wide-life, use of chemicals and pesticides, unimaginative mining water, ocean and air pollution, acidification, green house effect and ozone depletion. The efforts to improve the quality of environment and the issue of conservation and utilization of natural resources have also been delineated. Forests represent a well organised and highly evolved community of plants and animals. They provide several products of daily use such as food, timber, fire-wood, wood pulp, forage and fibre, apart from being a vast store house of medicinal plants which are yet to be fully explored and exploited. They are potent sources of many industrial raw materials. The greatest significance of forests, however, lies in their critical role in maintaining ecological processes and life support systems. According to official records 23 per cent of the land in India is classified as forests. The National Forest Policy of 1952 recommended that 33 per cent of the land area of the country should be under forests. Contrary to this guideline, however, forests have been under increasing assault since independence. It has been estimated that 1.3 million hectares of forests are lost every year. Land under forests shrunk from 46.42 million hectares in 1972-75 to 33.77 million hectares in 1980-82. At present not more than 10 to 12 per cent of the land is estimated to be under forest cover excessive exploitation of forests and overgrazing by cattle have seriously decimated our forest resource. The denudation of the Himalayan and other hilly area had led to soil erosion which affects river water quality, apart from inducing rapid siltation of dams and reservoirs. Degradation of forests leads to the destruction of wild life habitats. Over a hundred species of wild animals need immediate protection, as their populations have swindled to dangerously low levels.

Many plants have suffered from deforestation and destruction. In a recent list prepared by the Botanical Survey of india there are 135 species of plants that need immediate care for their continued survival. Deforestation also greatly increased the workload of local communities. Women and children, in particular, are forced to bear the burden of collecting fuel, water and food. A study in Almora reveals that because of indiscriminate deforestation the agricultural production, which was once sufficient, now sufficient, now feeds that villagers only for seven months in a year, as soil productivity has diminished and population has grown. Fire-woods and water have to be hauled from longer distances, because forests have shrunk and the springs in the neighborhood, once abounding in sparkling fresh water, have dried up. Pre-occupation with fire wood collection and hauling of drinking water leaves little time for the villagers to address themselves to other activities essential for socio-economic development improvement in their quality of life. Soil is the essential medium for the growth of plants. Management and improvement of soil fertility is vital for agricultural production and economic prosperity. But the pressure of the growing human and cattle populations has seriously affected the soil resources. The total cultivable land area in India is about 304 million hectares. According to an estimate made by the Ministry of Agriculture in march 1980, as much as 175 million hectares of this land area is suffering from environmental degradation.

Deforestation, overgrazing, unscientific agricultural practices and desertification have induced soil erosion. In Rajasthan, where only 20 per cent of the land is suitable for rainfed cropping, the mounting population pressure has doubled the area under cultivation from 30 per cent in 1951 to 60 per cent in 1971. This has happened mainly at the expense of grazing land necessary for pastoral agriculture which is the main occupation of a large section of the population in the state. With the population density already twenty times that of arid lands elsewhere in the world, the pressure of the growing population has led to various kinds of environmental implications.

More land being brought under cultivation, a decrease in grazing land, deforestation caused by rising demand for fuel and fodder, mounting pressure on land for house construction and other related developments have increased the desertification of western Rajasthan. This problem has been aggravated by wind and water erosion, water-logging, saline and alkali soils, primarily because of over-straining of the land resources. Soil degradation has serious economic and ecological implications for a country like india which has a prodiminently agricultural economy. For example, nutrient losses caused on account of NPK alone represents a loss of Rs. 700 crores per annum. Moreover, stream bank erosion and shifting of river courses through silt deposition by flood-prone systems affect the land use pattern. Soil erosion increases the silt load in a river system affecting aquatic productivity diversely and results in speedier siltation of dams and reservoirs. Siltation of dams reduced their effective life span, a factor which has serious economic and ecological implications.

The practice of cultivation, adopted in hilly areas covered with forest, has been a major cause of soil degradation. It is popularly known as the *Jhum* cultivation. It involves clearing of the forest by burning and using the cleared area for cultication. But on such fields cultivation is possible only for 3 to 4 years, after which the area is abandoned, as the soil fertility goes done. Thereafter a new forest area is cleared for cultivation. The abandoned area is used for regenerating the forest so that soil fertility is restored. In the past the cycle of this process used to take considerably longer duration. Since the population of the *Jhum* cultivator was small, it took about 600 to 1000 years of more to return to the same land for cultivation. But now a days the duration of the *Jhum* cycle has been drastically reduced, mainly because of the needs of the rapidly growing population on those areas.

Currently, only after every 4 to 10 years the same land is brought again under the *Jhum* cultivation. The reduced *Jhum* cycle is inadequate to recuperate the soil fertility. The continuation of the *jhum* cultivation, therefore, is increasingly promoting land degradation.On account of deforestation, tampering with nature, uncontrolled grazing in the catchment areas and soil erosion, floods have become a recurring phenomenon in many parts of the country. According to the report of the National Commission on Floods the total area affected by annual; floods had doubled since independence. While in the 1950s and 1960s three-fourths of all flood damage took place in five states, since the mid-seventies half the damage has been occurring outside the traditional flood zones. Loss of human lives and livestock as well as economic losses and damages to property are escalating steadily. The average economic loss due to floods during 1971-78 was nearly Rs. 1000 crore annually.

In the absence of proper land-use planning, floods have become increasingly devastating over the years. Moreover, the pressure of population growth force people to encroach upon flood plains. The disastrous impact of floods magnifies as the size of the population inhabiting flood plains and river banks increases. Ground water or aquifer is another major pool of fresh water that has been exploited by mankind from time immemorial. Ground water is periodically recharged by the rain water that seeps through the soil. It represents an important decentralized source of water supply which is available round the year. The total annual exploitable potential of ground water in the country is estimated at 42.3 m ha m (million hectare metre) of which only 10 m ha m is being exploited at present. Ground water exploitation for agriculture, industry and human consumption is rapidly increasing in many parts of the country to meet the demand of the growing population.

Large scale exploitation of ground water through dug wells and tube wells has created severe problems in many parts of the country. In many areas, ground water withdrawals have far exceeded the reaharging capacity of the aquifer, resulting in lowering of the water level and drying of wells. Intensively cropped one has undertake deep drilling. This calls for heavy expenditure, which is generally beyond the means of the average cultivators. This has resulted in chronic shortages of drinking water in many parts of Gujarat and Maharashtra.

Over-withdrawal of ground water in many coastal areas has promoted salt water or sea water intrusion, which has made the ground water unfit for drinking and irrigation. The use of ground water needs to be strictly regulated to prevent its population and over-exploitation. Flora and fauna are priceless fights of nature. They provide the essential underpinnings for cultural, industrial and economic development of a country. Nature has been very generous to india in providing a rich variety of flora and fauna which constitute its wildlife. The richness of Indian wildlife can easily compare with that of any other country of the world. More than 1200 species of birds and 500 species of mammals exist in the indian region and include such animals as the elephant, the gigantic himalayan sheep, the Indian bison or guar, deer, cheeta, the four-horned antelope, the black-buck, the dancing deer or thamin, the one-horned rhinoceros, the lion and the tiger.

It is however pertinent to note that populations of wildlife species are rapidly dwindling and many species,e.g. tiger and lion, once present in large numbers, are endangered today. Rapid increase in human and bovine populations has promoted large-scale habitat destruction by bringing more land under the plough and implementing developmental projects. This eventually has adverse impact on wildlife. To-day; the future of wild animals is, in general, alarming. Many endangered species are fighting a loosing a battle for survival. The worst victims, no doubt, are the fur-bearing mammals, birds with ornamental feathers and reptiles. Elephants are hunted for ivory and the rhinoceros is prized for its horn. About 350 species constitute the mammalian fauna of India, of which 81 are endangered. Examples of wildlife extinction in recent times are that of the Indian cheeta, the one-horned rhino and the Sikkim stag, and among the birds, the mountain quail, and probably among the timed species the Jerdon's courser. Over one hundred species of wild animal need immediate protection as their populations have swindled to dangerously low levels. Similarly, many plant species have suffered from the onslaught of human activities. Wild plants and animals are sources of many industrial raw materials in addition to providing a variety of other goods and services. It will be great tragedy, if wildlife is allowed to suffer from thoughtless destruction of their habitat by over exploitation, poaching or ignorance. The rapid pace of population growth demands increase in food production. Since independence, Indian has tripped its cereal production by adopting modern means of agriculture.

Chemicial fertilizers are the most essential input for modern agriculture. Fertilizer consumption has increased considerably. From almost zero level per hectare consumption in 1950-51, the consumption in 1986-87 was estimated to have gone up to 48.44 kg per hectare. India ranks as world's fourth largest producer of nitrogen fertilizer and the fifth largest producer of phosphatic fertilizer. At present the consumption of NPK in the country is 10.3 million tonnes. By the year 2000 the demand for fertilizers will go up to 20 million tonnes to help produce the projected food grain requirements of 240 million tonnes in order to feed the growing population. Such large input of chemical fertilizes will create diverse problems of environmental pollution which can be grouped into two categories.

1. Pollution caused during fertilizer production; and
2. Pollution resulting from the use of fertilizers

In the manufacture of fertilizers, large amounts of gaseous and liquid effluents are generated as by-products. The pollution cause by such effluents is relatively localized and can be remedied by reducing the volume of effluents or by treating them before their release into the environment. The Central and State Pollution Control Boards are persuading fertilizer manufacturers to cut down the release of effluents by enforcing Minimum National Effluent Standards prescribed for the fertilizer industry. The application of fertilizers in the field created many complex problems of environmental pollution.

A substantial portion of the fertilizers applied to the field is los through runoff, leaching, gasification or biological transformation. Runoff from agricultural fields laden with plant nutrients and excessive fertilizer water bodies is a condition technically called eutrophication. Over-loading water bodies with plant nutrients over-stimulates and disturbs the ecological Leakage of the ecosystems. Leakage of nitrate into the ground water creates the danger of nitrate poisoning. The problem of environmental pollution caused by the allocation of fertilizers in the field are widespread and difficult to control. Pollution loads build up in the environment in a field are widespread wand difficult to control. pollution loads build up in the environment in a cumulative manner, bypassing the existing pollution control legislations. The anticipated intensification of agriculture in the future is going to magnify the problems of environmental pollution in the coming decades.

Pesticides are a group of chemicals used for controlling pests, insects, weeds, disease-causing bacteria, fungi and viruses. The organochlorine compounds, which include insecticides like adlrin, dieldrin. DDT and its derivatives, are causing great concern to environmentalists. DDT is highly resistant and gets widely dispersed throughout the ecosystem. It is present in rainfall, soil and in wildlife dispersed throughout deer and the Arctic penguin. It is soluble in fat and hence, tends to accumulate in very large quantities in the liver of animals occupying the higher levels in the food chain, e.g., man, birds of prey,etc. The use of pesticides in India has grown rapidly since independence DDT and BHC are the two main pesticides used in large quantities, DDT has been very effective against insect pests that destroy crops and insect vectors transmit malaria and plague. Since the cropped area under plant protection has increased from 6.4 million hectares in 1960-61 to over 80 million hatchets in 1980, the average consumption of pesticides increased from 3.2 g/ha in 1954-55 to 336 g/ha in 1980. The use of these agrochemicals is going to increase in future. The success of pesticides has been a mixed blessing because their use has serious environmental implications. In 1962 Rahcal Carson, in her famous book, *Silent Spring*, highlighted the dangers of introducing large quantities of non-specific pesticides.

A comparative evaluation of DDT in human fat tissue from different countries shockingly revealed that residents in Delhi had very high levels of DDT, DDT and BHC have been found to be abundantly present in samples of butter and wheat flour collected from Punjab. Even mother's milk has been found to be contaminated with DDT. Both DDT and BHC have a lot of residue toxicity and can even cause cancer. Pesticide use therefore should be strictly controlled and regulated. With a view to controlling the sale and use of pesticides, the government of India passed an Insecticide Act in 1968, but it has been poorly implemented. Mining is an extractive process. Coal, lignite, iron-ire, limestone and bauxite are mainly obtained for surface mining. The Indian mining industry is expanding at a steady rate of around 5 percent. The value of mineral production increased from Rs. 85 crore in 1951 to Rs, 9953 crore in the year 1987. Mining operations induce serious hydrological disturbances, as some of the aquifers may be intercepted by mine openings.

The underground hydrological system is greatly disturbed. Dewatering of mine pits causes water pollution of streams and reservoirs. The Damodar river in Bhowrah has turned black in colour due to suspended solid discharged by the coal washeries in Jharia. Water of many streams in the Bihar-Orrisa iron-ore belt is redish yellow due to the discharge of iron-ore fine and lateritic slimes. Also water pollution sometimes occurs due to breaches or overflowing of tailing ponds. The influx of labour engaged in mining creates problems of water pollution due to the discharge of raw sewage. Dust generated by mining and blasting operations, by the movement of heavy earth-moving equipment and by over-burdened dumps, contributes to air pollution. Diesel operated equipments, such as shovel, dumper, tractor,etc. release lots of Co, SO_2, No_x and aldehydes into the environment. Mining and mineral processing have a great impact on the environment. Open-cast mining is highly devastating and leads to large scale environmental degradation. Limestone mining in th Mussoorie-Dehradum region and bauxite mining by the Bharat Alumnimum Company in the Gandhmardan hills in Orrisa, have led to widespread public outcry against the environmental degradation.

The main environmental problems caused by open-cast or surface mining are: deforestation, wildlife destruction, land damage, hydrological disturbances, water and air pollution, noise and ground vibrations, displacement of population, socio-economic changes and aesthetic blight. Although the actual mining is carried out in a relatively small area, its impact spills over to a much larger area. In open-cast mining the mineral is obtained by removing the overlying non-ore material. Huge piles of over-burden keep growing around the mine pit creating a 'hill and dale' topography. Over-burden and digout pits are abandoned after the mining excavation is over, since there is no legal obligation foe reclaiming the area after mining is over. Mining brings in additional traffic load in the area, leading to the road congestion as well as increase in noise levels. Moreover, eviction of local inhabitants from the ore-bearing area leads to serious socio-economic distortions. The phenomenon of environmental refugees as a result of such environmental degradation poses serious problems. The ancillary and down-stream industries, drawn by the mining activities, case further deterioration of the environment. Water is the life-blood of the environment. Without water no living thing, plant, animal or man can survive. The availability of an adequate and usable supply of water underpins our whole economy. Water is used for transportation and power generation,for waste disposal, recreation, agriculture, and fisheries. It is essential both for the manufacturing and the service sectors. Requirements of fresh water for different uses during 1974 and the estimated figures for the years 2000 and 2025 are give in the following Table.

Table 1 Water use of major activities in India, 1974-2025.

Use	1974	2000	2025
Irrigation	350	630	770
Domestic & Livestock	13	34	50
Industries	5	30	120
Thermal power generation	10	60	160

It may be seen that by the beginning of the twenty first century, there will be a six times increase in the water required for industry, and by the year 2025 it will be 24 times the present demand. A rational pragmatic strategy for the management and recycling of water needed to cope with such galloping demands. The problem of great concern. Water pollution is the most important cause of public health problems, both in rural and urban areas.

The frequent eruption of epidemics are caused by the pollution of drinking water. Waterborne diseases, such as cholera, typhoid, diarrhoea, dysentery, malaria, and intestinal worms claim a heavy toll every year. Inadequate hygiene, poor sanitation and discharge of industrial effluents have resulted in the pollution of many of our surface water sources. The main sources of water pollution are summarized as: (1) urban/domestic effluents; (2) industrial liquid effluents; (3) surface runoff from cultivated fields where fertilizers, pesticides, insecticides and other agro-chemicals are used; (4) surface runoff from urban and industrial solid-waste dumping areas; and (5) nuclear and thermal plant discharges.

Domestic wastes constitute approximately 90 per cent of total pollution load, while 7 per cent from large and medium industries and 3 per cent from small scale industries. In Delhi, domestic wastes constitute approximately 98 per cent by volume. In Bombay, the industrial capital of india, it is approximately 84 per cent. The Sabarmati is the most polluted river of the country with a BOD (Biological Oxygen Demant) level of 1655 mg/1. The Mahi, the Narmada, the Tapti, the Krishna, the Cauvery, the Pennar, the Godavari and the Subarnarekha rivers are also polluted in stretches. All the major rivers immediately downstream of big cities are polluted. The Ganga river is most seriously polluted near Kanpur, Varanasi and Howrah. The BOD levels in Varanasi and Kanpur are at times reported to be higher than 30 mg/1.

The status of sewage control and treatment in Class I cities and in Class II cities is very precarious. India's progress in controlling domestic wastes, as surveyed during 1978-80 is indicated. As regards water pollution caused by industries. 17 categories of industries have been identified as contributing to it. Out of the 1.700 large and medium scale industries identified as water polluters, approximately 800 have established pollution abatement devices.

Table 2. Status of disposal of domestic waste water

Class of City or Town	*Number*	*Sewered (Percent)*	*Unsewered*	*Waste water volume (millions of liters/day) Generated*	*collected*	*Treated*
Class I	142	43	57	7.000	598	378
Class II	198	12	88	1,200	158	58

Ocean and coastal areas have also suffered from the pressure of growing human population. They are being used as convenient dumping sites of all kinds of water, including toxic and hazardous materials, without giving much thought to the consequences. In addition, the use of sea routes for transporting oil cargo causes problem of oil pollution of the marine environment. One of the important oil tanker routes passes through the Arabian Sea and Southern Bay of Bengal and it lies close to our western sea coast. Leakages and oil spills from tankers are the prime causes of oil pollution of the marine environment. Resides, when tankers are involved in accidents, the impact on ocean water is still more serious. Crude oil, being immiscible and lighter than water, quickly spreads and floats over the sea surface. Oil slicks are carried to the shore by sea currents. An oil cover on the sea surface blocks the free exchange of gases between sea water and the atmosphere.

Floating oil slicks chocke the gills of fish and clog the plumage of marine birds. In short, the communities of marine organisms are completely destroyed. Offshore drilling and oil production yet another potential threat for scale oil pollution of the marine environment. Recent satellite surveys suggest that the oil pollution of the Arabian Sea has increased considerably. Beaches along the west coast of India are plagued with tar-ball deposition which adversely affects their use as recreation spots and discourages tourism. Clean-up operations are very expensive, cumbersome, inefficient and are rarely carried out. Consequently, oil pollution of the marine environment generates several complex problems. Air pollution is one of the major problems. It is caused by moving and point sources. Moving sources are represented by automobiles, ships, aircrafts and rockets. Industries and thermal power plants are examples of point sources of air pollution. The burning of fossil fuel products oxides of sulphur, nitrogen carbon dioxide, carbon monoxide and a complex mixture of organic compounds along with particulate matter. The most important pollutants in the Indian environment are suspended particulate matter and So_2.

The SPM levels in most of the urban areas are beyond the permissible level. In fact, an unpleasant pall of dust has become almost an integral part of urban areas. Air pollution problems are most visible. They affect living organisms significantly. Apart from creating problems of poor visibility, dust pollution causes respiratory diseases particularly in children and elderly persons. Sulphur dioxide is the most important among the gaseous pollutants. It is released by thermal power plants and other industries along with a large amount of flyash. The problem of flyash and SPM pollution caused by power plants, cement and other industries, can be controlled by their installing electrostatic precipitators. The gaseous pollutants, however, escape into the atmosphere. Growing consumption of fossil fuels has enormous implications for the air quality. An estimate made some time ago indicates that SO_2 emission from fossil fuels ins the country has increased from 1.4 tonnes in 1964 to 3.2 m tonnes in 1979. It is expected to increase further as the tempo of industrial development gains momentum. Sulphur dioxide affects all living organisms, but plant are particularly vulnerable to this gas. The average content of lead in the environment is increasing. Tera-ethy lead is added in petrol as an anti-knock agent. A study carried out in Delhi has revealed that the lead content in the ambient air near the road side varied between 90 to 324 narogarms per cubic meter, while its value never exceeded 62ng/m in the background station located away from vehicular traffic. Acid rain, which has emerged as the great scourge in the industrial countries, is regarded an imminent threat to India, where the emission of acidic gases is sharply increasing. In the last few decades the rain water has become acidic over large areas in Europe and North America. The term 'Acid Rain' is used to describe all precipitation rain, snow, sleet, dew which are more acidic than normal. Acid deposition is caused when sulphur and nitrogen oxides from coal and oil fired power plants, from industrial processes, such as metal smelting, and from motor vehicle exhausts, combine in the atmosphere with water vapour, sunlight and oxygen to form sulphuric and nitric acids.

Environmental acidification, is thus, a man-made phenomenon. Emission of sulphur dioxide and oxide of nitrogen are prime contributors to atmospheric acidification, as they readily dissolve in the atmospheric moisture forming sulphuric and nitric acids which make the rain acidic. The ecological impacts of acid precipitation are far-reaching, insidious and often produce irreversible changes. Acid rain has a corroding effect on plant leaves and thus pre-disposes them to pests, pathogens and other environmental stresses. The soil becomes acidic-a condition which releases toxic heavy metal ions in the soil and promotes the loss of calcium and magnesium from the soil. the cumulative effect is the gradual degradation of soil and a decline in forest productivity. In aquatic systems, acidic rain raises the levels of toxic heavy metals resulting in the death of fish and the reproductive failure. Increased acidity of lakes and streams prevents hatching of fish eggs and kills they young fry. Acidification in industrialized countries is responsible for making thousands of lakes unproductive, turning them into virtual biological deserts. The vast expansion of thermal power generation and other industrial activities contribute to environmental acidification. The release of sulphur dioxide and oxides of nitrogen are rapidly increasing in the Indian environment and this trend is going to intensify in the future due to growing demands of our galloping population. Systematic and reliable data on the acidity of rain water In india are lacking, but occasionally acid rain has been reported from Bombay, Delhi, Nagpur, Pune and a few other station.

Carbon dioxide is a natural constituent of the atmosphere and has concentration of a little over 0.032 per cent by volume, a ratio of 1 to 450 with oxygen. in spite of its relatively small proportion, carbon dioxide plays a very important and useful role in the biosphere until its ratio is disturbed. It is like glass, transparent to visible light but absorbing the infra-red radiations. It lets the sun's rays through to the earth, but traps the heat that would otherwise be radiated back into space. While the natural levels of carbon dioxide keep the earth's temperature at a comfortable 15 C. any increase in its level makes the earth hotter. It is in view of this phenomenon that a great concern is growing because of the steadily increasing carbon dioxide concentration in the atmosphere. It is estimated that from 1700 to 1975 the concentration of carbon dioxide in the atmosphere has increased from 270 ppm to 334 ppm. Its over-all concentration has increased by nearly twenty five per cent since the process of industrialisation started in the world. Most scientists are therefore in agreement that global warming is underway. This phenomenon is known as the 'Green House Effect". Besides carbon dioxide, other trace gases like methane, nitrous oxide, the chloro-fluoro-carbons and ozone also produce the Green house Effect and have an additive action. What will be the environmental impact of this phenomenon? How will plants, animals or ecosystems as a whole respond to it? Precise predictions are difficult, but significant studies indicate that die to a global increase in carbon dioxide, the temperature of the lower atmosphere is likely to increase by 1.5 C to 45 C by the year 2030. This would lead to a rise in the global sea level because of two major reasons. First, the rise in atmospheric temperature will make the average temperature of the oceans warmer. Because of this heat transfer, the water of the earth will expand. Second, the rising temperature will result in the melting of polar ice caps, which will ultimately result in the rise of the sea level.

Empirical studies predict that global warming would lead to a rise in sea level between 1.4 metres to 2.2 meters by the end of this century. This would have obvious implications to millions of people living in coastal areas and delta regions of rivers like the Ganges. Increasing carbon dioxide will induce sweeping changes in the biosphere. Plants will be affected by the rising levels of 'green house gases' in two ways. First, the high concentration of carbon dioxide will have a fertilising effect and will encourage plant growth. Second, the concomitant increase in the ambient temperature will include long0-term climate change. Climate shift will upset the present day cropping pattern over large northward. Insects will breed faster. Many wetlands will dry out, some food chains will be altered and broken or some will disappear. the over-all impact will be distressing. Ozone is a deep blue gas made up of chemically bonded oxygen atoms and is a minor constituent of the earth's atmosphere. It is found everywhere between sea level and a height of 60km in varying concentrations. In the air we breathe, ozone is a health hazard, a constituent of air pollution that has a caustic effect on human skin. However, in the stratosphere, ozone forms and delicate veil, filtering out UV radiation from sun's rays entering the earth's atmosphere. In fact the presence of an ozone layer in the stratosphere is vital for life on earth because it is the only natural shield against UV radiation, which is a potent mutagenic agent. It has recently, been discovered that the protective ozone layer is getting progressively eroded due to the impact of increasing human interference with the environment. A significant reduction in ozone layer over Antarctica is already a matter of concern. The major cause for the depletion of the ozone layer is the worldwide emission of man-made compounds called choloro-fluoro-carbon.

Chloro-fluoro-carbons are used in refrigerators, air conditioners, acrosols, sprays, for the cleaning of computer chips, and in the making of rubber foam and polystyrene containers required for food packaging. CFCs are by and large chemically inert, having no direct effect on humans or other lying organisms. After their release they ultimately find their way to the stratosphere where they chemically react with ozone molecules, breaking them down. If the release of CFCs continues at the current rate, the ozone layer will be further depleted and living organisms, including human beings, will be exposed to higher levels of UV radiation. Scientists believe that increased doses of UV radiation will cause eyes damage, skin cancer, and will accelerate the aging process both in humans and domestic stock at all latitudes. The seriousness of ozone depletion is similar to the hazards of nuclear disaster. It is urgently necessary to prevent further damage to the ozone layer.

In India there has been a growing realisation of the impact of the increasing population and accelerated development strategies on the quality of the environment. The increasing human demands have been exerting great pressure on the natural resource base, agricultural land, energy sources, vital watersheds, forests, etc... resulting in environmental degradation. Recurring natural calamities, such as floods and droughts, have attracted the attention of one and all.

The concern was well reflected in the Parliament when a specific provision on protection and improvement of the environment and safeguarding of forests and wildlife was added to the chapter on directive Principles of State Policy by the Constitution Act, 1976. Article 48 A of this chapter states: 'The State shall endeavour to protect and improve the environment and to safeguard the forests and wildlife of the country.' This constitutional amendment further evidenced the concern of the Members of Parliament for environmental improvement by incorporating Part IV A on Fundamental Duties which provides for a duty of citizens to the environment. In its Article 51 A Constitution states: 'It shall be the duty of every citizen of india...(g) to protect and improve the natural environment including forests, lakes, rivers and wildlife, and to have compassion for living creatures.

However, the official concern for the environment was expressed for the first time in the Fourth five Year Plan Document, which emphasised the need to introduce environmental dimension into the development planning process. In 1972 a Committee on environmental Planning and Coordination was set up to advise the government on all environmental matters. In January 1980 another Committee was set up to suggest legislative measures and an administrative machinery for ensuring environmental protection.

On the recommendations of this Committee, the Department of Environment was set up by the Government. Subsequently, a separate Ministry of Environment and Forests was established in 1985 to deal with various environmental matters. the government has initiated a number of measures to protect and upgrade the environment.

Thirty major enactments related to the protection of the environment are now being administered by the central and state governments. The Water Act, 1974 and the Air Act 1981, the Factories Act, and the Insecticide Act are some of the prominent ones among these enactments. The Bhopal gas tragedy in 1984 came as a rude shock and it exposed the dangers of hazardous chemicals. A Union Carbide pesticide plant accidentally released a cloud of deadly methy I isocyanate over the town. This accident killed about 2500 and more than 200,00 people fled for their lives.

Nearly 100,000 people are still suffering from various kinds of side effects, such an blurred vision, lung diseases, intestinal bleeding and neurological disorders. This led the Government to enact the Environment Protection Act to deal with the environmental matters in a comprehensive manner and to plug the loopholes in the existing legislations. The governments has made it mandatory that environmental issues and unintended ecological side effects of the proposed development projects are identified in advance and necessary corrective measures are incorporate to reduce the undesirable environmental impact.

Project approval by the Ministry of Environment and Forests often requires implementation of suggested safeguards and mitigative measures. The Central and State pollution control Boards have been constituted. The Central Board has prescribed minimum national standards for liquid and gaseous effluents in respect of 12 major polluting industries. An ongoing programme of monitoring and assessment of air and water quality is being carried out to keep a close watch on the emerging trends. There are a number of critical environmental challenges that are being met by undertaking suitable measures. Soon after independence in 1974, the Government of India took note of the extensive destruction of the forest wealth of the country. In order to reverse this trend and to arouse public consciousness about the importance of forest in the welfare of the nation, an annual tree planting festival was started in 1950. Since then each year, tree planting campaigns have been undertaken while celebrating 'Van Mohotsava'. In the early years there was a great enthusiasm and the response of the people was quite encouraging.

Unfortunately, over the years it has lost its vigour reducing it to a mere ritual without much impact. The concept of 'Van Mahotsava' is highly relevant but unfortunately it has entered into deep dormancy. Deforestation has been occurring for many years and about half of the area under forests sin the country is degraded. Ecological restoration of damanged areas is needed for meeting the socio-economic needs of the people. The National Agriculture commission recommended that 'Social Forestry' be undertaken to meet the fuel-wood, fodder and small timber needs to rural communities. The main objectives of social forestry are to relieve the pressure on existing forests,s to promote the productive use of public and common lands in a decentralized manner, to promote soil and water conservation and to create employment opportunities in rural areas. It envisages creation of local village level nurseries and involvement of the local population including women and children to raise trees of their choice to meet their needs.

This includes the planting of trees on denuded and even on agricultural lands inter-mixed with crop plants or along farm boundaries or on vacant lands. Depending on local conditions, it can take any one of the following emphases: Agri-silvicltural, Agri-pastoral or Agri-silvipastoral. The government has been laying considerable empphasis on social forestry programmes since 1986 and fund allocation for this purpose has been steadily growing. Tree seedlings are being distributed free to the poor people in rural states to give usufruct rights over trees planted and protected on public lands by the rural poor.

The usufruct rights provide the individuals the right to use the trees and get profit out of them, but they cannot be regarded as their property. Until recently, tree planting and afforestation programmes have remained mostly confined to quick-growing trees like Eucalyptus or the commercial species. This has increased the raw material supply instead of mitigating shortages of fuel-wood and fodder. To bring about qualitative changes in the programme, in June, 1988, the national waste-lands Development Board was set up with the principal aim of regenerating wasteland through massive afforestation and tree planting. Waste-land maps have been prepared for 146 selected districts in 19 states, showing village-wise distribution of 13 categories of Wastelands. Village-specific action plans identifying land treatments, if any, need to be prepared.

The future programme would focus on drylands for the production of fuel-wood, fodder and small timber on public lands an multiple purpose trees under farm forestry. The Government of india announced a comprehensive National Forest Policy in December 1988. The principal aim of the new policy is to ensure environmental stability and equilibrium, which are vital for sustenance of all life forms, human, animal and plant.

The derivation of direct economic benefits must be subordinated to this principal aim. A great deal of organisation effort will be needed for the successful implementation of this policy. Enlightened public opinion and active involvement of people is necessary for protecting and improving the local environment. A local leader, Chandiprashad Bhatt started a practical and non-violent methods—the 'Chipko Movement'—to save the forest of Uttarakhand from the axes of contractors. On 27 March, 1973 the residents of Mandal village of Chamoli district lid not allow as sports goods manufacturing firm of Allahabad to fell the sash trees auctioned to them by the U.P. government.

The sports company moved towards Mandakini Valley of the Kedarnath region. But there too they met with similar resistance. Hill women and children showed remarkable courage and joined the 'Chipko Movement' and successfully save the village trees from felling. The 'Chipko Movement' highlighted the value of conservation over the extraction of temporary economic benefits trees. a new slogan was given, namely 'What do the forests bear?. Soil, water and pure air. Soil, water and pure air are the basis of life. ' A massive planting of trees was started to provide the Five 'Fs'-food, fodder, fuel, fertilizer and fiber. Within a few years, the 'Chipko Movement' developed into a powerful mass- based ecological movement in the country.

Various animal species and plants are great assets for the future well-being of society. Wild plants and animals have provided breeding materials for improving the commercially useful species. A few instances will display their invaluable contribution to human welfare and prosperity. A wild melon from India provided the genes for resistance to mildew of a melon crop as far away as in California in the united States. A wild rice collected in 1963 in Uttar Pradesh gave the genes which saved 30 million hectares of paddy from grassy sweet virus. The *'Kaus'* grass from Indonesia provided the genes for resistance to redrot disease of sugarcane. Among animals, the showl cows of Punjab and Haryana and the goat of Etah district of Uttar Pradesh have been utilised for improving the breeds of these animals in many countries. *Penicillium,* a microscopic of soil fungus, which is a source of the popular antibiotic Penicillin and *Cinchona,* from which quinine is obtained, were unnoticed species until their useful properties were discovered. What a great loss it would have been if these two species had become extinct before the discovery of their medicinal value.

Our knowledge about the majority of wild plants and animals is very limited, and hence loosing them without knowing their virtue will be a great tragedy. But deforesation, poaching, monoculture, pollution and habitat destruction have already caused extinction of many species of plants and animals. The widespread use of selected high-yielding varieties of crop plants as well as mechanisation of agriculture had dangerously eroded the genetic diversity of agriculturally viable species. According to an estimate, the number of endangered species of flowering plants in the country has sharply risen from hundreds to thousands over the past few years. It is no feared that 10 to 20 per cent i.e. over 2500 species of the total vascular plants, may now fall in one or the other categories to threatened species.

All species, whether they are economically useful or not and whether their economic uses are known or not, are crucial importance in the ecosystem. Long term measures directed towards conservation are necessary for protecting the plants and animals. The conservation of threatened and endangered wildlife species is best achieved when their natural habitats are protected. In our country a number of natural areas have been set aside with a view to conserve various endangered wildlife species. There are now sixty-seven National Parks and 394 Sanctuaries distributed all over the country, representing 4 per cent of the total geographical area. They cover a wide range of ecosystems ranging from forests, grasslands, deserts, river basins, lakes, wetlands, mangroves and marine areas. Special efforts have been made to conserve the magnificent tiger, which is our national animal, by establishing 17 tiger reserves, covering an area of 26,634 sq. km. in 13 states. It is now increasingly recognised that conservation of wild areas along with the inherent diversity of their plants, animals, insects and micro-organisms is vital for sustainable development.

Long-term conservation of the representative ecosystems of the country has been initiated by creating four Biosphere Reserves namely, the Nilgiris, Nanda Devi, Nokrok and Great Nicobar. Another nine potential biosphere sites have been identified. Almost all facets of human activity produce large quantities of undesirable by products which are called wastes. For every ton of grain, fruit or vegetable produced, between tow and three tonnes of solid waste are generated. Based on a 'global average', it has been calculated that a city of one million inhabitants generates 500,000 metric tonnes of waste water, 2,00 metric tonnes of solid wastes and 950 metric tonnes of air pollutants every day. The quality and variety of wastes are growing with urban and industrial expansion. Waste disposal is becoming a major concern demanding a substantial share from the budgets of civic bodies. Waste disposal, being non-remunerative, is usually treated with neglect, resulting in the piling up of waste heaps in towns and cities resulting in frequent outbreaks of epidemics. This need not be so, if waste recycling is practiced and by-product utilization is taken up seriously.

Nature does it for itself all the time. In sharp contrast to human societies, recyling is a norm in nature, hence the problem of waste is non-existent. For example, green plants utilize carbon dioxide produced by other living organisms or released from the decomposition of dead plants and animals. Wastes are of two types, namely organic and inorganic. They can be potential sources of materials or energy or both, provided they are imaginatively recycled. Wastes recycling and utilization can take any one or more of th following forms, namely retrieval of useful material, resource separation, energy recovery and innovative utilization. Waste picking is an established practice in urban and industrial areas. Useful, materials such as glass bottles, containers and metal scrap are recovered and sold.

A 10 MW capacity thermal power plant is already operating using rice husk in Punjab. Similarly bagasse is another potential source of energy. A 3.75 MW plant has been installed din Delhi to generate power from the incineration of the city waste. It has been convincingly demonstrated that cattle dung, human excreta, agricultural and domestic wastes are suitable feed-stock for biogas generation. In the Okhla area of New Delhi the city sewage is utilized for generating biogas which is supplied to the nearby colonies ad domestic fuel. Large quantities of flyash produced in coal-fired thermal power plants can be used for variety of purposes such as for the production of cement and bricks, for applying as soleing-material for roads and air strips and for using as building material. Waste recycling and reuse are essential for resource conservation, environmental.

Waste recycling and reuse are essential for resource conservation, environmental improvement and sustainable development. The following examples indicate the patterns of the implications involved in the recycling process. Considerable amount of agricultural and animal wastes are traditionally composted and used as farm yard manure in villages. But with the advent of inorganic fertilizers, this practice if fading way over the years. In almost all towns and cities, the most common method of disposing solid waste is by using it for land filling and land reclamation. In big cities like Calcutta, Bombay, Delhi and Madras several residential flats have been constructed by reclaiming land through this process. This is seldom carried out scientifically and hence the land fill sites become potentially dangerous to public health.

The Ganges, one of the most celebrated and revered rivers in the world, is also one of the most polluted. Almost one-third of the country's population lives in the Ganga basin. Out of 100 towns and cities situated along its banks, 20 per cent represent major urban conglomerations. The rapid growth of these centres has led to over-exploitation of the rivers, leading to rapid degradation of the river water quality. Some 600 km of its total length are dangerously polluted with human, animal, agricultural and industrial wastes.

The majority of villages, towns and cities suited on the banks of the Ganga have no sewage treatment plants and, even if they exist, they are far too small, badly maintained and poorly operated. Frequently water flows through open drains into the Ganga. Experts say that if sewage discharge alone could be stopped, pollution levels of the river would drop by 75 percent. Contamination of water by sewage is the principal cause of water-borne diseases, including cholera, typhoid and para-typhoid fever, dysentery and infectious hepatitis. In addition to sewage, dead bodies, half cremated corpses, dead animals and synthetic detergents are additional factors responsible for destroying the health of the river Ganga. Detergents release large amounts of phosphates, which over-fertilize the river water, leading to many undesirable effects on the water quality. The practice of intensive agriculture in the Ganga basin has greatly promoted the consumption of chemical fertilizers. The Ganga is constantly fed by surface run-off from cultivated lands which have been treated with excessive amounts of fertilizer and manure. A two-year study conducted by the Central Pollution Control Board indicates that 1.15 m tonnes of pesticides were applied in the Ganga Basin- an amount equal to 34 per cent of India's annual consumption. A large proportion of this ultimately finds its way to the Ganga through run-off waters. Recognizing the need to protect human health and the ecology of the river, the government of India has launched a major clean-up programme, called the Ganga Action Plan. Over the next five years, the government has commuted Rs. 250 crores for the rehabilitation of the river. During the first phase of the action plan, a number of measures have been undertaken.

A number of projects are being implemented to renovate the existing sewage systems in all towns and cities along the Ganga to upgrade all existing sewage treatment and pumping plants, to install new sewage treatment plants wherever necessary and to extend sewage line into areas not currently covered by the treatment network. These projects also relate to the construction of large community cattle sheds in urban areas to collect animal wastes for biogas and fertilizer and regulation of the use and application of pesticides and chemical fertilizers in agriculture in the watershed of the Ganga to minimise surface run-off.

One of the important features of the Ganga Action Plan, therefore, is to recover and recycle the organic wastes for useful proposes. Sewage is a very rich source of energy and of organic manure. A proper sewage treatment plant can yield both energy as biogas and slurry which can be used as an organic manure. The interrelationships between population, resources and environment are very complex. It demands a consummate understanding for appreciating the linkages that are crucial for sustainable development. However, it is universally accepted that human survival depends on resources like air, soil, water, forest, wildlife and minerals, which nature has bestowed on us. It is also being increasingly realised that rapid population growth has resulted in great pressure on natural resources. Let us examine the potentials and limitations of our natural resources, which can be grouped into *living* and *non-living* resources. They can also be categroised on the basis of their inherent characteristics into two categories, namely *renewable* and *non-renewable* resources.

Some resources are inexhaustible and keep on renewing themselves if managed properly. These are known as renewable resources. But resources like coal, petroleum, etc. are non-renewable, as the stock of these resources is fixed and with constant use their quantities decrease. The regeneration capacity of even renewable resources is seriously impaired, or many even be lost, if these are not properly managed.

Extinction of plant and animal species, desertification, soil erosion and salinisation are examples of renewable resources becoming non-renewable through overuse and excessive exploitation. The natural reserves of metals and minerals are fixed and finite. Strictly speaking they are non- destructible and can be recycled repeatedly. But in practice they get dissipated and dispersed through human use. Sometimes metals are used any once metal is cumbersome or expensive. It is therefore very much desirable that natural resources are conserved and managed wisely to avoid their wastage or under utilization. The conservation and efficient utilization of living and non-living resources is vital for meeting the demands of the growing population.

Mineral resources greatly contribute to the generation of wealth. Our country is richly endowed with minerals. India is self-sufficient in 35 minerals which are used as raw materials by various industries. Table 6.3 gives an idea of the available quantities of some important minerals in the country. The main mineral producing states are Bihar, Madhya Pradesh, West Bengal, Gujarat, Rajashthan and Adndra Pradesh. The mining and mineral industry has steadily grown during the past years. The mineral production in 1976 was worth Rs. 1,365 crore and it rose to Rs. 9,641 crore in 1986. The share of fuel minerals, non-metallic minerals and metallic minerals was worth Rs. 8,226, Rs. 770 and Rs. 644 crore respectively. The major mining operations in india are open-cast, which severely affect the land-use pattern in the area. Land subsidence in underground mining areas affects the general landscape very adversely.

The total coal resources in India according to the estimate made by the geological Survey of India is 117 billion tonnes. Out of this, 86,427 million tonnes can be categorised under reserves. These consist of coal seams situated upto 600 metres of depth and more than 1.2 meters in thickness. About 5.094 million tonnes of coal deposits can be categorised under conditional resources which cannot be mined under present-day technological options. The balance resources are distributed between speculative and hypothetical resources depending upon the degree of certainty in the exploration of deposits.

Table 3 Estimated reserves of some important mineral resources in India

Mineral	*Estimated Reserves in Crore tonnes*
Bauxite	265.37
Chromite	13.53
Copper Ore	57.60

Diamond	10.00
Dolomite	395.00
Gold	148.51
Fire Clay	49.28
Gypsum	124.86
Graphite	46.41
Iron Ore	19.757
Lead Sin	35.85
Limestone	7.320
Manganese	13.50
Nickel Ore	23.13
Tungstan	71.00
Refrctory Minerals:	
(Magnesite, Kyanite, sillimanite)	23.91

Estimated Reserves in lakh tonnes

The proven coal reserves in the country are of the order of 156 billion tonnes. The actual coal production in the country has increased from 55.67 million tonnes in 1960-61 to 180 million tonnes in 1988-89. By 2004-04 the coal production is expected to increase up to 450 million tonnes per annum. Almost two-thirds of this will be from open-cast mines and one-third will be from underground mines. The balance recoverable reserves of oil and natural gas in the country are at present estimated at 58.1 crore tonnes and 541 billion cubic metres respectively. The annual production of crude oil was around 320 lakh tonnes in 1987 and that of natural gas was bout 981.2 crore cubic meters in 1986-87.

The balance recoverably reserves of natural gas increased from 35,200 crore cubic meters as on January, 1980 to 54,100 crore cubic meter in January 1987. The present requirement of petroleum products is in excess of domestic availability and the shortfall of about 6 to 17 million tonnes is being met through imports. The demand for petroleum products is expected to rise to about 111 million tonnes by the year 2004-05, whereas the domestic production is likely to attain a level of only about 58 million tonnes. We will, therefore, have to continue to depend on imports to meet a substantIal part of our demand for petroleum products in the coming years. The position in respect of the availability of natural gas is, however, expected to improve considerably over the coming years and the reserves of natural gas in the country are expected to last much longer than those of crude oil. In 1987 there were 12 refineries in the country having a total capacity of 467.00 lakh tonnes. Power is the most basic requirement for industrial and economic development. The total energy consumption of 186.3 million tonnes of coal replacement in 1953-54 has grown to about 750 million tonnes of coal requirement in 1988 and is expected to rise to 1224 million tonnes by the year 2000. The power generation scenario shows that thermal power is most important modality followed by hydro-electric power, while nuclear power contributes only about per cent to the total power output. In the foreseeable future, the contribution of thermal power is going to grow further while the relative position of hydro-electric and nuclear power generation will remain the same.

Thermal power generation has a wide variety of direct and indirect effects on the environment. The direct effects are due to the combustion of coal which gives rise to a number of air-borne effluents, the most important of which are sulphur oxide, nitrogen oxide, carbon monoxide, carbon dioxide, particulates, organic compounds, trace metal and radionuclides. Substantial amounts of nitrogen oxides, are generally present in the effluent gases. The source of nitrogen oxides is not the fuel *per se* but the nitrogen in the air combines with the oxygen in the air during the combustion process to produce NO. Later, most of the NO oxides to NO_2, which in turn, can combine with water to form nitric acid aernsol. Hydro-electric generation has a number of environmental impacts. No dam can be built and no impoundment can be created without affecting the environment. A dam becomes a dominant factor in the hydrological regime, and sets in motion a series of impacts on physical, biological and socio-cultural systems. The dam and the lake behind it cause many effects on the environment regardless of the dam's geographical location.

The environmental side-effects of dam construction are generally divided into two categories: (a) the local effects and the reactions within the area of the man-made lake, and (b) the downstream effect resulting from a change in the hydraulic regime. Both types of side-effects of hydro-electric generation have a complex impact on the physical, bilogical and socio-economic environment: The folding of the region could have immediate and significant impact on the means of communication, historic sites, communities which are inundated, and the local flora and fauna. In this connection the Silent valley represents and excellent example. A hydro-electric dam was proposed to be constructed in the Palghat district of kerala. The project involved the construction of an RCC arch dam of a georage formed by two hillocks rising approximately 160 to 250 meters above the bed level. The dam was supposed to submerge over 670 hectares of virgin tropical rain forest, which to the project proponents was not very important.

Ecologically, the area represents the highest expression of mother nature and there are hardly and obstacle, not only to the free running of water, but also to fish migration and evolutionary work of nature, and in view of the overwhelming ecological considerations, the silent valley project was dropped altogether.

Environmental hazards of nuclear power have been the subject of heated debate the world over. The environmental consequences of nuclear power generation are most complex and potentially dangerous. The health hazards of radiation are the consequences of alpha and beta particles and gamma rays emitted by radioactive materials. These particles and rays can permanently change the hereditary materials of cells in living organisms, thereby inducing cancer. Firm evidence suggests that people who have been exposed to low level radiation over extended periods of time show an increased incidence of leukemia, skin, thyroid and lung cancer. For example, studies of children who have received X-ray treatments in the neck region have shown an increased incidence of thyroid cancer.

Generally, a five-to-twenty-year latency period elapses following exposure before these diseases are diagnosed. Low level exposure also accelerates the ageing process, thus shortening life span, and can cause chromosimic damage. The human foetus is particularly susceptible to radiation damage. Exposure to radiation has an additive, or cumulative effect. The radiation from medical and dental X-ray are added to the natural background radiation, thus, resulting in greater total exposure. The half lives of radioactive wastes produced by fission reactors vary widely from 0.96 second for xenon-143, to long as 24,400 years for plutonium-239. Usually at least 10 half-life periods must elapse before a source decays to a point at which it no longer constitutes a serious radiation threat. The nuclear waste must be kept in isolation to prevent human exposure. The first several hundred years are most critical.

Future generations will have to watch over these wastes for centuries. The accident of the Three Mile Island nuclear power plant at Middle Town, Pennsylvania, U.S.A. on March 28, 1979 and the

Chernobyl accident in the Ukraine in the USSR on April 26, 1986 have provided sharp focus on the hazards of nuclear power. The important environmental hazards of nuclear power are summarised below:

1. Mining of uranium-radon gas in uranium mines is a major health hazard;
2. Enrichment of fissionable uranium from 0.7 per cent to about 3 per cent;
3. Pelletization of uranium'
4. Transportation of radioactive materials;
5. Leekage of radioactive material from fuel element to receiving body through cool-out;
6. Disposal of nuclear waste material-currently about 2700 tonnes of spent nuclear fuel are stored at sixty five nuclear power plants throughout the country.
7. Human error and/ or mechanical failure internal to installation;
8. Human error and/ or mechanical failure external to installation;
9. Natural catastrophe; and
10. Malicious human activity

These and other considerations suggest that the chance a major nuclear accident may not be vanishingly small and the environmental consequences are truly serious. The energy that conventional sources can provide is far below our present requirements. In addition, the conventional sources of energy like firewood and fossil fuel are fast shrinking. During many year in the past we have suffered form fossil fuel are fast shirking. During many years in the past we have suffered from shortages of energy mainly because the energy demand jumps ahead of its supply.

Therefore, to meet the growing energy demand of the expanding population, new and renewable energy sources need to be developed. In 1982, a separate Department of Non-Convention energy sources was created by the Government of India to promote the harnessing of solar, wind, biomass, geothermal oceanic energy sources.

Solar energy, unlike other conventional energy sources, is received abundantly all over the country and provides a source of inexhaustible energy which is non-polluting. India is indeed fortunate to receive an abundance of sun0shine with about 164802100 Kwh/M^2/ year with 250 to 300 days of useful sun-shine a year. There are a number of possible ways to use in abundance the renewable supply of solar energy.

The important among them are various types of heating devices, such as, solar coolers, solar dryers and solar heaters, which use the heating property of solar energy. solar energy can also be converted into electricity by photovoltaic systems. This is accomplished by a special silicon cell which produces electricity when exposed to sun-light. The output from one such cell is small, but a panel of silicon cells connected together produces sufficient power to run water pumps, traffic lights, radio, TV or other appliances.

Wind is a renewable source of energy and one of the earliest sources of power exploited by mankind to augment his muscle power. The windmill is an invention that sates from the earliest times of recorded history. Egyptians used windmills as early a 3600 B.C. to lift water. The use of wind energy for driving ships, lifting water grain grinding, sugarcane crushing, turning of the machines of factories, and for doing many other task developed progressively over the centuries. In the nineteenth century wind was used to generate electricity.

Although wind is ubiquitous, it varies widely from one station to another in its intensity and magnitude. It also varies diurnally and seasonally. Therefore, its potential differs drastically from one location to another. Windmills can be installed in those parts of the country where an average wind speed of 8-10 kmph prevails. Coastal areas, mountain tops, grasslands, and islands are the more favourable sites for harnessing wind power. Since wind does not blow all the time, the power output of wind turbines is highly variable and sometimes unpredictable. In India, a number of water pumping windmills have been indigenously developed. Over 1.750 pumping windmills had been installed under a demonstration project by March 1987. Five wind farms, four of 5450 K and one of 1.1 MW, installed in the costal areas, have contributed over 4.0 million units of electricity to the national grid. The relative contribution of wind power in comparison with other energy sources is going to be small. However, this sources could make vital contributions in remote areas with favourable wind resources. The Department of Non-conventional Energy Sources has taken the initiative for wind monitoring and preparation of wind maps of different parts of the country.

Mankind from time immemorial has been using biomass to meet the energy needs. Biomass includes, strictly speaking, all living material, whether plant or animal. Interests has focused recently on new and more efficient sources of biomass creation and improved methods or processing to retain ad much energy as possible and produce usable fuel. Cattle dung, human excreta and other organic wastes which have biological origin are unusable biomass and pose a grave danger to public health, if not disposed off safely. The biogas technology which has developed during the past few decades has opened up the possibility of disposing these obnoxious wastes by converting them into useful products, mainly biogas and organic manure. A biogas plant operates in the simple principle that when cattle dung or any organic residue is fermented under anaerobic conditions or in the absence of free oxygen, a combustible gas in generated. Agricultural, domestic, human and animal wastes, water hyacinth or any other organic material can be used as feed-stock for biogas production. Distillery wastes can be suitably used for biogas generation. Obnoxious aquatic weeds, like water hyacinth, can also be used as feed-sock for biogas generation. The gas produced by the bacterial feeding on the organic material consists of two parts of methane and one part of carbon dioxides. It is clean pollution-free fuel. It is tapped from the gas collector and can be used for cooling, lighting, pumping water, electricity generation or as fuel for any other purpose. The gas production is maintained by periodically introducing fresh dung into the digester. The spent slurry over-flown from the digester is relatively free from pathogens and is extremely suitable as organic manure or soil conditioner. In rural areas, biogas offers an excellent solution to the potential sources of environmental degradation.

Organic wastes which are potential sources of environmental degradation are converted into useful products namely clean fuel and organic manuar, much needed by rural communities. Biogas offers many indirect advantages such as conservation of forests which are being increasingly exploited for firewood by the rural communities. It can also relieve women and children from the drudgey of collecting and transporting had loads of firewood over long distances. Village sanitation can also be improved if latrines are attached to biogas plants. China and india are world leaders in biogas technology, The central government encourages installation of biogas plants by providing cash subsides.

By March 1987 over 4.8 lakh family type biogas plants were already installed in our country. Geothermal energy, the natural heat contained within the earth is generally too deeply buried to be utilized. However, in some areas of the world that have experienced recent volcanic activity, geothermal resources may exist that can be exploited economically. These are usually found in regions of hot springs or areas where the deep hot rock is known to be fractured or to have spaces that permit water or steam to be circulated to carry heat to the surface. At such places source for electricity generation, space heating drying and refrigeration. It is competitive with other technologies.

In our country the potential for geothermal energy has been identified as occurring in 4 regions: (1) the Non-Western Himalayas covering Ladakh. Himachal Pradesh, Punjab and Uttar Pradesh, (2) the Narmada-Sone Valley.(3) the Damodar Valley, and (4) the West Coast. Forty six hydrothermal areas have been identified where the temperature of the spring water exceeds 150C. A. 7.5 tonne capacity cold storage pilot plant based on geothermal energy has been installed at Manikaran in Himachal Pradesh as a joint venture of the Indian Institute of Technology. Delhi and the Geological Survey of India, Lucknow. Plans ark afoot to use geothermal energy at Pauge Valley, Ladakh for poultry framing and mushroom cultivation. The form of energy available from the oceans include Ocean Thermal Energy Conversion, wave energy and tidal energy. The surface layers of the oceans are continually heated by the sub- especially in the tropical belt. Enormous quantities of energy ar stored in the surface layers which have higher temperatures ranging between 24 C to 30C. However, at lower depth, 1,00 M or below, the water temperature ranges between 4 C to 10C. This temperature difference is exploited to generate electricity by a process called Ocean Thermal

Energy Conversion. The technical feasibility of OTEC is well established, but the costs are high and further development work needs to be done before it is harnessed. The potential can be successfully exploited for generating electricity. The surging and receding tides produce powerful water currents in run electric turbines for electricity generation. Even the most slowly reproducing species would cover the earth in a short time if its population growth were unrestrained. Malthus understood that this "implies a strong and constantly operating check on population from the difficulty of substance. Pushing the point further he wrote: Through the animal and vegetable kingdoms, nature has scattered the seeds of life abroad with the most profuse and liberal hand. She has been comparatively spring in the room and the nourishment necessary to rear them. The germs of existence contained in this spot with ample food, and ample room to expand in, would fill millions of words in the course of a few thousand years.

Necessity, that emperious all pervading law of nature, restrains them within the prescribed bounds. The race of plants, and the race of animals shrink under this great restrictive law. Darwin echoed this view in On the origin of Species.... as more individuals and produced than can possibly survive, three must in every case be a struggle of existence, either one individual with another of the same species or with the individuals of distinct species, or with the physical conditions of life. It is the doctrine of malthus applied with manifold force to the whole animal and vegetable kingdoms, for in this case there can be no artificial increase of food and prudential restraint from marriage. Although some species may be now increasing, more or less rapidly, in numbers, all cannot do so, for the world would not hold them. At about the same time Herbert Spencer, an English philosopher and contemporary of Darwin presented a more rigorous, quantitative concept of population regulation in his book First Principles. Every species of plant and animals prerpetually undergoing a rhythmical variation number now from abundance of food and absence of enemies rising above its average, and then by a consequent scarcity of food and abundance of enemies being depressed below its average....mid these oscillations produced by their conflict, lies that average number of the species at which its expansive tendency is in equilibrium with surrounding repressive tendencies. Now can it be questioned that this balancing of the preservative and destructive force which we see going on in every race must necessarily go on. Since increase of numbers cannot but continue until it is either arrested by fertility or extinguishes the race entirely.

This essentially modern view of the regulation of population grew out of an awareness of the immense capacity of populations for exponential increase. In a sense, a populations, a growth potential and the relative constancy of its numbers cannot be logically reconciled otherwise. However, to say that faltering birth rates and increasing death rates check a populations as its numbers grow is to do little more than elaborate the definition of a regulated population. Yet during the 1920s, when Lotka and others gave population biology its quantitative foundations, the projection of growth in a regulated population was one of the first problems to be tackled. The phenomenal population growth of the American colonies had greatly impressed upon Malthus how rapidly humans could multiply; this was not so evident in the more crowded European countries of his time. Pearl and Reed wished to project the further growth of the population, which they supposed must eventually reach a limit. Data for the population to 1910, the latest census then available, had revealed a decline in the exponential rate of growth.

Pearl and reed reasoned that if the decline followed a regular pattern that could be described mathematically then one could predict the future course of the population, so long as the decline in the exponential growth rate continued. They also reasoned that changes in the exponential rate of growth must be related to the size of the population rather than to time because the time scale is arbitrary with respect to any particular population.

Malthus's arguments about population has stimulated considerable interest in mathematical formulations populations growth and many scholars had taken into account that imperious, all pervanding restrictive law of necessity which restrains populations within prescribed bounds. The most successful of these attempts was by the mathematician Pierre-Fraanois Verhulst, who, in 1838, published an equation identical to that developed by Pearl and Reed 82 years later. In a paper, Verhulst named this formula the logistic equation because of its logarithmic exponential form.

Many scientists of the day discussed the equation and compared it with less successful alternatives, but Verhulst's contribution to population biology was submerged by a wave of other scientific developments, particularly, and somewhat ironicially, that of Darwinism; the logistic equation disappeared from view by mid-century.

The early twentieth century saw increasing interest in mathematical biology and the accumulation of data to which calculations could be applied. In 1908, T.B. Robertson independently Verhulst equation to describe the growth of organisms. Robertson contained the name "autocatakinetic" for his equation, capturing the essence of his idea that populations of cells, like populations off individuals, increase by self replication but eventually come under the influence of growth restrains. In 1911 A.G.M Kendrik and M.K. Pai, working at the Pasteur Institute of Southern India. Again rederived the logistic equation to describe the growth of population of the gut bacterium *Esherishia coli* cultured in artificial medium. In 1913, the Swedish biologist Tor Carlson devised some wonderful equations to describe the growth of cultured year cells. One of these had the from.

$$dN/dt = rK^{n}e^{qt}(1-N/K)^{n}$$

Which made *r(N)* both an exponential function of time (*t*) and a function of population size (*N*). The equation had a variable, *r*(*N*), to rescale the effect of populations size on *r* in a nonlinear fashion. Then in 1920, Pearl and Reed, apparently unaware of these earlier papers wrote their equation for the growth of populations.

The following year, Pearl acknowledged Verhulsts work and his name logistic for their equation. The logistic, finally made generally known was the first equation in ecology to generate research that would provide the data to test a mathematical model. Pearl reviewed the growth of populations in an article published in the *Quarterly Review of Biology* 1927.

Besides data on human populations and carlson's excellent data on yeast, which pearl have a prominent place, the only experimental work to which he could refer was his own on the fruit fly *Drosophila*. This was followed quickly, however, by similar work on flour beetles by Chapman, water fleas by Tera and Tenaka and the protozoan *Paramecium* by Gause they all attempted to determine whether the growth of populations fit the shape of the logistic curve. By and large they were successful, and the logistic equation took a permanent place in population biology.

Lotka showed the relationship of the logistic equation to Taylor's polynominal series

With his characteristic mathematical insight, Lotka pointed out that the logistic equation is merely the smallest number of terms of a Taylor expansion that can satisfactorily characterise restrained population growth Taylor's that any equation $f(X)$ in a single variable X can be approximated by series of terms of the form $a + bX + cX^2 + dX^3 + ...$ Lotka pointed out that any equation describing the growth rate of a population in terms of its size can therefore be expressed.

$$dN/dt = a + bN + cN^2 + dN^3 + ...$$

Applied to the problem of describing population growth the first term (a) must be deleted because as population size goes to zero the realities of life require that population growth rate also goes to zero. The second term is simply exponential growth, with $r = b$; hence it alone is not sufficient to describe regulated population growth. The second and third terms of Taylor's series together are, however, adequate if b is positive and c is negative; hence $dN/dt = bN - cN^2$. Rearranged this equation becomes $dN/dt = bN\ (1 - [c/b]N)$. Substituting r for b and k for b/c gives the familiar logistic equation. There is no reason why additional terms of Taylor's series should not be added to an equation for population growth. Carlson realized, the logistic cannot be beat for simplicity and adequacy. Lotka was quick to point out that blind application of equations like the logistic was meaningless and potentially misleading unless one understood the biological processes responsible for the behaviour that one could describe mathematically.

The logistic equation describes a pattern of population growth; its terms do not explicitly describe underlying processes. Clearly, biologists needed to delve more deeply into the factors that caused logistic like population growth curves to crop up time after time. In their monograph on the growth and control of the gypsy month population in the United states, L.O. Howard and W.F. Fiske made the useful distinction between factors whose effect on the population depended on density and those whose effect did not. In order this balance may exist it is necessary that among the factors which work together in restricting the multiplication of the species there shall be a least one, if not more, which is what is termed facultative and which by exerting a restraining influence which is relatively more effective when other conditions favor undue increase, serves to prevent it.

A very large proportion of the controlling agencies, such as the destruction wrought by storm low or high temperature, or other climatic conditions, is to be classed as catastrophic, since are wholly independent in their activities upon whether the insect which accidentally suffers is rare or a abundant. The storm which destroy 10 caterpillars out of 50 which chance to be upon a tree would doubtless have destroyed 20 had there been 100 present, or 100 had there been 500 present. The average percent age of destruction remains the same, no matter how abundant or how near to extinction the insect may have become. Howard and Fiske's "facultative" and "catastrophic" factors were called "density-dependent "and density-independent" factors by H.S. Smith and these names have stuck.

Many things influence rate of population growth, but only density-dependent factors, whose effect varies in relation under to crowding can possibly bring the population under control. Of prime importance among these must be limitation by food supply and place to live and predator parasites and diseases whose effects are more strongly felt in Crowded than is sparse populations. Other factors such as temperature and moisture, influence birth and death rates without regard to numbers of individuals in a population; they therefore contribute only to the linear term of Taylor's series that of exponential growth, while not influencing the quadratic term, which expresses the force of regulation. Pearls experimental studies on *Drosophila* provided he earliest indication of how density affects population processes. when confined to a bottle with a fixed supply of food, the descendants of a single pair of flies increase in number rapidly at first, but soon reach a limit. To measure the effects of density on birth and death rates, pearl introduced different numbers of pairs of files to otherwise identical culture bottles and recorded deaths of adults and number of progeny produced per day.

Pearl found that the number of progeny raised per pair decreased as he density of flies in the bottle increased. Subsequent work has shown that this effect expresses competition among the larvae for food, which results in high mortality in dense cultures. pearl also showed that adult life was cut short at high densities, well above the levels that affected the survival of larvae. Pearl's work provided the beginnings of what has become a vast body of research on the detailed relations of populations processes to the density of individuals. Indicative of the early progress in this field was the work of A. C. Crombie, of the University of Cambridge, on the grain beetle rhizopertha dominica, Rhizopertha is a tiny beetle that completes its larval development within a single grain, living off the kernel of the seed.

Chapman and others had already determined that grain beetle populations follow logistic growth curves, implying density-dependent regulation. Females lay their eggs on the surface of seeds. Immediately after hatching the larva bores its way into the seed, where it commences development; later, the mature larva bores its way out of the same seed o pupate. Though seed coats reduce the success of first instar Rhizopertha larvae entering grains of wheat, and mortality at this stage may reach 50 per cent. Once inside, the larva enjoys security so long as it is alone; a kernel of wheat cannot support more than one beetle. Yet, in a dense population, females may lay many eggs on the same grain. Crombie discovered that females lay eggs without regard to the presence or absence of other eggs on a grain of wheat. He the sought to determine whether the larvae bored into grains at random or avoided those already occupied. To do so, he placed 150 wheat grains in a dish and introduced 100 first instar larvae.

What happens when two larva meet in the same grain is another story, however. As crombie described it: When two larvae in the first, second, third or fourth,instars were put together into a small hole drilled in a wheat grain and watched under a binocular microscope, they were often seen to attack each other with their mandibles and eventually either one or both left the hole when a larva entered such a hole it always went to the bottom and turned round so as to face outwards.

Other larva trying to enter the hole were fiercely attacked. Sometimes such combats resulted in the body wall of one of the antagonist becoming punctured and its bleeding to death. In their tunnies in wheat grains larvae of all instars were always found curled up with the head facing towards the way they had entered. Furthermore, in all grains dissected during the experiments to be described, whenever two larvae were found in the same tunnel at least one of them was a always dead.

The experiments Crombie referred to we re as follows. He infested a single wheat grain with from 1 to 8 larvae, giving them 6 hours to become established the grain. He then transferred the infested grain to a dish with 5 other fresh grains and left them together for 48 hours. At the end of the experiment, he dissected all the grains to determine how many larve had been killed and how many had migrated to a different grain. The results showed conclusively the strong effects of density on the survival and movement of larvae even within the first 2 days. Crombies beetles contested each grain of heat through direct confrontation. After all, sole possession of a single grains is a larva's ticket to success. In this population, density dependence came into play abruptly as the number regardless of the number of larvae of larvae relative to wheat grains the end result was always the same one grain one larva.

Unlike grain beetles, water fleas exert a less direct influence on each other through their consumption of resources. each water flea consumer millions or prey the single celled green and diatoms of the plankton. As prey are consumed the available the food to other water fleas decreases gradually leading to a graded response of birth and death rates to prey density. to demonstrate this effect P.W. Frank, C.D. Boll, and R.W. Kelly cultured water fleas in small beakers in which populations were maintained at densities between 1 and 32 individuals per cubic centimeter they added green algae periodically to each culture, and noted the survival and fecundity o females for 2 months after the beginning of the experiment.

Life tables for populations at each density revealed that fecundity decreased markedly with increasing population at each density. Somewhat unexpectedly, survival increased at densities up to 8 individuals cm^{-3} before decreasing at higher densities. At densities of 8 individual cm^{-3} and above, stunted body, growth suggested that depletion of food resources between periodic replenishment of the alga stock culture limited birth rates and survival; the effect was clearly density dependent.

The geometric rate of population growth (λ) calculated from the water flea life tables decreases linearly with increasing density and falls below 1.0 at a density of about 20 cm^{-3}. Therefore, under the conditions of temperature, light, water quality, and food a availability provided in the laboratory, *Daphnia* populations would attain a stable size of about 20 cm^{-3}, regardless of the initial density of the culture. The data of Frank and his colleagues on water fleas relates birth and death rates to population density, but does not elucidate the mechanism of density dependence. Depletion of food seems the most likely explanation for reduced survival and fecundity at increased density, but excreted waste products, such as ammonia—which may accumulate in confined populations to the point that they act as poisons could also cause density-dependent effects.

Stephen Carpenter attempted to disentangle the effects of food and waste products in an experimental study of density dependence in the mosquito *Aedes riseriatus*, a species native to North America that carries human encephalitis in some areas. In the upper midwestern states, water filled holes in trees provide particularly favorable breeding sites; mosquito larvae graze the protozoa, bacteria, and fungi that grow upon decaying leaves and other plant detritus in the three holes Carpenter reproduced the essential character of these breeding sites in the laboratory by partly filling plastic petri dishes with water retrieved from the trunks of trees rainstorms and providing measured quantities of beech leaves as a natural substrate for the microorganisms the mosquito larvae would feed upon.

To distinguish the effects of food supply and water products, he established cultures of larvae at all 27 combinations of three populations sizes and three volumes of water. To detect the biological effects of these treatments, Carpenter measured the survival of the larvae and the mass of the pupae; the latter partly determines the fecundity of the adult female. If food alone caused density dependence these traits would respond only to the ration of leaves per individual independently of the number of individuals or the volume and pupal mass would respond to the number of individuals per volume of water. The concentration of ammonia at the end of the experiment varied as predicted with respect to each of the factors: greatest at the lowest volume of water, at the highest number of individuals within each volume, and at the highest ration for each number within each volume. The effect of ration on ammonia suggest that the metabolism of the mosquito larve varies in direct relation to the amount of leaves, presumably a consequence of higher feeding rates. The response of the mosquitoes to each of the factors clearly demonstrates the overriding importance of food ration on survival and pupal mass, and equally clearly sows that the levels of excretory products obtained in the experiment were not high enough to inhibit growth and survival. Within each volume treatment, and for each number of individuals per treatment, both survival and mass increased markedly with food ration.

Density dependence influences reproduction in population of deer

Most studies of density dependence have centered upon laboratory populations where factors can be controlled experimentally. But the simplicity of such systems also leads one to question the relevance of laboratory findings of populations in more complex natural surroundings-where physical conditions changes continuously, and food and predation are not controlled by the experimenter. Ideally one would like conduct the same experiment in nature as in the laboratory; that is, to alter the density of individuals in the populations while keeping all else constant. In practice, this difficult experiment can be accomplished only with populations managed intensively for some other purpose. Game animals are sometimes maintained at altered levels by management practices, and ecologists have taken advantage of such situations to study population processes. Studies of various species of deer illustrate the expression of density-dependent factors. The reproduction and survival of deer depend directly upon the quality of the food supply. Deer browse leaves, and require large quantities of new growth with high nutritional content to maintain high growth rates and normal reproduction. The direct physiological effects of food availability upon reproduction were shown in a study of white-tailed deer in New York state where the proportion of females pregnant and the average number of embryos per pregnant female were directly related to range conditions.

The number of corpora lutea in the ovary indicates the number of eggs ovulated, hence the reproductive potential of the female. the difference between number of corpora luetea and number of embryos showed that poor range condition, resulting in poor nutrition of pregnant females caused embryo death and resorption. In the central Adirondack area, where the habitat for deer was very bad, even ovulation was greatly reduced. Many studies on deer have shown that high populations can lead to deterioration of the habitat over time though overgrazing. In one population in Ohio, whose size had increased following protection from hunting deteriorating range condition caused by overgrazing led to decrease in growth rate, adult body weight ovulation rate, and production of fawns. Range deterioration caused by overgrasing often can be reversed by selective hunting to thin dense populations. During the course of Chaetum and severinghau's study of the white-tailed deer in New York, one of the word ranges in the state located in the Adirondack Mountains, was opened to hunting. Two study areas, DeBar Mountain and Moose River, had been investigated between 1939 and 1943, just prior to the open hunting seasons of 1943 and 1945, and then again in 19947. Because Moose River is relatively remote, few deer were shot; but in the more readily accessible De Bar Mountain area, 250 deer were removed from the 8000 acre range in 1945 alone. And reproduction improved dramatically over its level before the hunting and in the relatively unchanged Moose River area.

In crowded populations, plants suffer high mortality slowed growth and reduced fecundity

The sensitivity of life values of plant to density has been shown in many studies, summarized in detail by the British plant ecologist John Harper. Like animals plants experience increased mortality and reduced fecundity at high density. Unlike many animals however, plants respond to density by slowed growth, along with its consequences for fecundity and, to a lesser extent survival. Starve an animal and most often you will kill it, Starve a plant and it won't grow. This difference arises in part from the modular construction of plants.

The repeated units of leaves and buds that make up plant body exist more or less independently of each other; that a branch fails to grow has little effect on the well-being of other branches on the tree. The more complicated arrangement and interdependence of the parts of animals dictate greater constraint on size and shape. Flexibility of plant growth is revealed in the size of flax plants grown to maturity at different densities. When sown sparsely at a density of 60 seeds per square meter, the modal dry weight of individuals was between 0.5 and 1 gram, and many plants at trained weights exceeding 1.5 g. When sown at densities of 1440 and 3600 seeds m^{-1}, most of the individuals weighted less than 0.5 g, and few grew to large size.

The variation in size within a planting arises due to chance factors early in the seedling stages, particularly the date of germination and quality of the site in which the seedling grows. Early germination in favorable spot gives the plant an initial growth advantage over others, which becomes accenturated as the larger plants grow and crows their smaller neighbors. The flexibility of plant growth does not preclude mortality in crowded situations. When horseweed seed was sown at a density of 100,000 m the young plants completed vigorously. As the seedlings grew, many died and the density of surviving plants exceeded the decline of the population, and the total weight of planting increased.

Over the entire growing season the hundredfold decreases in population density was more than balanced by a thousand fold increase in the average weight of each plant. When one plots the logarithm of average plant weight as a function of the logarithm of density, data points recorded during the growing season fall on a line with a slope of approximately—3/2. Plant ecologist call this relations between average plant weight and density a self thinning curve. Such is the regularity of this relationship that many have referred to it as "the—3/2 power law', although is biological basis is not understood. The self thinning curve, whose position on the graph varies from species to species, appears to represent the maximum average dry weight for each density of a population.

When seeds are planted at a moderate density, so that the beginning combination of density and average dry weight lies below the self thinning curve, plants grow without appreciable mortality until the population reaches it self thinning curve. After that point the intense crowding associated with further increase in average plant size causes the death of smaller individuals.

Stunted growth under crowded dramatically alters the life tables of plants. Palmblad germinated seeds of the plantain plantago major at densities of 1 to 200 per pot. Although only quarter of the plants in the densest treatment died many of the individuals that flowed failed to produce seeds, eventually lacking the resources to do so. Although the total dry weight of all the plants in each pot taken together was similar across densities the total number of seeds produced decreased from about 12,000 per pot a low density to 4420 at the highest density.

The year 1954 brought the publication of two important books on population biology, each remarkable on its own merits, but even more striking together in the utter opposition of their points of view. The English ornithologist David Lack, in The natural Regulation of animal numbers, vigorously advocated the regulation of population size by density-dependent factors, principally food, predators and disease. His arguments and evidence added to the momentum of belief in density dependence and the regulation of populations about stable equilibria by Lotka Gause, A.J. Nicholoson and others.

The same year, two Australian entomologists, H.G. Andrewartha and L.C. Birch, published. The Distribution and Abundance of Animals in which they argued that most populations particularly those of insects and other small invertebrates are influenced primarily by density independent factors and that periods of a favourable environmental conditions for populations growth ultimately control size of the populations. The numbers of animals in a natural population may be limited in three ways: (a) by shortage material resources such as food, places in which to make nests etc.; (b) by inaccessibility of these material resources relative to the animals capacity for dispersal and searching; and by shortage of time when the rate of increase is positive.

Of these three ways the first is probably the least and the last is probably the most important in nature. Concerning c, the fluctuations in the value of may be caused by weather, predators or any other component of environment which influences the rate of increase. Although Andrewartha and Birch marshaled abundant evidence to support their view that exponential growth rates reflect density independent factors, they were greatly influenced in their thinking by the study of J. David son and Andrewartha on populations of the tiny insect pest Thrips imagines near Adealaide, South Australia. The tharip which infests roses and other cultivated plants undergoes regular periods of increase when seasonally favourable conditions prevail, followed by rapid decline when conditions are less suitable. Adeladie has a Mediterranean climate winters are cool and rainy, summers not and dry.

The spring brings an ideal combination of moisture, warmth, and plant flowering. Thrips subsist mainly on plant pollen, whose abundance varies seasonally with the production of flowers. But Adelaide has a sufficiently mild climate that some flowers are always available, and thrips remain active all year. During the winter, however, the depressing effect of cool temperature on development rate and fecundity causes the thrip population to decline to low levels. Population growth is further checked by high mortality of immature stagesthrips take so long to mature during the winter that before many can complete their development, the flowers in which they live have withered and fallen to the ground carrying the thrips with them to their deaths. The warm whether of sporing speeds development and increases the daily fecundity of females.

5 Population Growth

Population studies are fundamental to many human problems in agriculture, pest control, wildlife management, and economic planning. The purpose of population research is to understand the interplay of factors influencing population range, and ultimately to predict and regulate the course of populations. Man obviously wants to limit or reduce populations of organisms, whereas he has wish to maximize within reasonable limits valuable and productive animals for aesthetic or economic reasons. The greatest challenge of all, as well as scientifically and philosophically the most difficult are of human endeavor is understanding and regulation human population change.

The world population was almost stationary during Stone Age, an estimated 10 million people. Sometime between 8000 B.C. and 6000B.C. man learnt the art of growing food, and so could support a large population. As a result the human population increased 50 times in 7000 to 9000 years. In the next two centuries the population doubled and reached an amazingly high figure of one billion. It took just 100 years to double the population to reach four billion in 1975. Alarmingly, in 12 years it reached the five-billion mark on July 11, 1987. At the present rate of growth we will add another billion in the next 12 years, unless drastic measures are taken. The present world population of 5.43 billion, which is growing at the rate of 97 million people per year will touch 8.5 billion by the year 2025. About 95% of the population growth will be in the developing countries. At present we add about 500 million people in about six years times, a figure which took thousand of years to reach before 1950 AD. About 3,42,000 babies are born every day in the world and about 1,5,000 persons die, leaving a net increase of 2,07,000. Thus within a week's time there will be nearly 1.5 million more people. The UN population Fund in a recent report says that in the next 10years there will be one billion more people on Earth, making a total of 6.4 billion. That means another China to house, a million more consumers to feed, house, clothes, and educate.

There is no doubt in the fact that human population are drastically out of balance. For thousands of years prior to the seventeenth century, the population of the world was less than 500 million people. Human population were limited by high mortality and even in the great civilizations of Egypt, Greece and Rome, expectation of life probably did not exceed 30 years. The class checks on human population were disease, famine and war. In the fourteenth century, at least 25 percent of the adult population of Europe died in epidemics of bubonic plague and between the years 1348 and 1379 A.D. England's population was reduced almost 50 percent, Cholera, typhus, small pox, malaria yellow fever and sleeping sickness have all been among the great killers of man throughout history. Several of these diseases, especially malaria and yellow fever, severely limited the abundance and distribution of people in tropical countries until the advent of modern medicine and public health. Asia has traditionally been the great stage for famine though starvation; death also occurred in Europe. The famous Irish famine of 1850 resulted in the death of more than 12 percent of the adult population between 1846 and 1851. In China, in 2000 years preceding the 20th Century, over 1800 separate famines were recorded, many of them killing million of individuals. More recently, 4 million famine deaths occurred in China in 1920-21 more than 5 million is Russia from 1918 to 1923; and over 2 million in India in 1943. Almost agriculture and transportation have alleviated famine mortality, malnutrition continue to be significant demographic for, particularly since it accentuates illness and mortality from infectious disease. War has also been a major mortality factor throughout history,not only by causing direct battle casualties, but also by stimulating disease and famine. Great epidemics and massive crop failures, have often been triggered by the disruptive condition of warfare. The casualties of war have frequently involved as many civilians as soldiers. Some historians believe that the Thirty year war in Europe between 1618 and 1648 resulted in the death of 30 percent of the inhabitants of Germany and Bohemia.

Despite the combined action of disease, famine and warfare, several major development in the sixteenth, seventeenth and eighteenth centuries stimulated population growth and eased the heavy mortality burdens of man. The opening of the New World provided a vast new area of space and nature resources of economic exploitation. The industrial revolution greatly increased the productivity and control of infectious diseases. The industrial revolution greatly increased the productivity and mobility of man. These factors combined to accelerate human population growth and the world population curve began to rise upward. Between the birth of Christ and the seventeenth century, the world's population took approximately 1600 years to double. The next doubling as achieved within 200 years, and the next from 1 to 6 million people within 80 years, from 20 to 4 million people were completed by 1975 a span of 45 years and at the current rates of growth the world population will double again within another 35 years. Currently the world's population is increasing at approximately 2 percent per year—a net increase of about 1,300,000 per week. Every day of every week, the world's population increases by more the 207,000 people. All this is primarily a product of man's increased life span a reduced mortality through modern medicine and agriculture. Man's fertility patterns were basically established during periods of much higher mortality rates thousands of years ago and they have remained high in the intervening year.

The developing countries least able to afford rapid population growths are experiencing the greatest growth. The population is not evenly distributed around the world. More than half the world's population regions of the total world, i.e. Europe, North America, former USSR and Oceania had only about 22% of the total world population in mid 1988. The remainder was distributed in the developing region of the world. The growth rate is most rapid in Africa. In the coming decade, the population there will rise from 640 million to 950 million and it will be the highest regional growth rates the world has ever known. According to projection, countries like Ethiopia, Nigeria, Zaora and Kenya will double their populations within the next 20 years. In North Africa the growth rates are slower, but th figures are still rising fast. South Asia and Africa together will account for over half of the increase. Southern Asia's population will rise from 1.2 to 1.6 million. In Eastern Asia the population growth rates have declined rapidly in the past decade, but the number will still increase from 1.3 to 1.5 million people. Several Arab countries have very high population growth rates and the 12 countries in West Asia will increase in their population to 131 million by 2001. The population; of Latin America and the Caribean is projected to increase by 100 million from 440 to 540 million. In Central and South America, most countries have annual growth rates in excess of 2%.

In addition to the large concentration of population in developing countries, the percentage of population under 15 years and over 64 years of age what is called the dependency ratio is also very high. The population under 15 year is about 45% in Africa, 35% in Asia and 39% in latin America, whereas it is only about 23% in North America 22% in Europe and 29% in Oceania. However, the percentage of population over 64 year is higher in the developed regions of the world. North America has about 11% of its population over 64 years. Europe has about 13% and Oceania about 8%. It is only 3% in Africa and 4% in Asia and latin America.

The high dependency ratio in the developing countries means that age group between 15-64 has to support a large population with requirements for housing, clothing, education, health and other social service. The large young population in developing countries also has demographic implications in the near future. Asia and Africa, and the regions which will account for about 75% of the world's population in the next 35 years or so, and will have the greatest proportion of poor people.

In South Asia 51% of the population lives below the poverty line and in sub Saharan Africa 45% Malnutrition, ill-health, illiteracy and the lowest life expectancy, accompanied by inadequate technology and economic poverty, are predominant in South Asia and Africa. It is not very difficult to imagine the miserable plight of the people in their regions in the next century. The UNEP has warned that continuing high population growth will result in continuing poverty and increasing environment damage and will lead to higher levels of urban growth and international migration. The population report support this that it will be impossible to keep pace with population trends. If the population growth continues as expected, it says within 10 years there will be more than 200 million children who will receive no education. What does this mean in terms of human welfare and environment quality? It usually means that many countries have excessive demands for products and services which cannot be met. It is exceedingly difficult, if not impossible, for those countries with the highest rates of population growth to meet the requirements of their people for food, housing, jobs and the physical amenities of life, and it is virtually impossible for them to maintain adequate services in education, medicine, public health and social welfare.

Thus we emerge with such problems: (1) 60 percent of the world's people are still inadequately fed, receiving a percapita calorie, intake less than that recommended by the UN (2) 80 percent of the children of India shows evidence of malnutrition; (3) there are only 9 acres of land surface available for every person on the earth, and only 26 percent of this is available for agricultural production; thus only 1.2 acres of potential cropland are currently available for each person and this is decreasing each year (UN States of Hunger Report); (4) many countries, especially in Asia and Africa, still have an illiteracy rate exceeding 70 percent; (5) in many developing nations, there is only one physician for every 5,00 to 10,00 people, a ratio which makes adequate medical care impossible. This list could be extended considerably to indicate that many parts of the world are currently living under severe population pressure. While the western nation gain strikingly in prosperity, most of the world's people face increasing poverty and population pressure.

When to global picture is considered, the outstanding fact exists that we can not provide a quality life or maintain a quality environment for the majority of the world's people at the present time, let alone provide for more than 1,600,000 new people every week. Ecologically, this represents a highly unstable condition, it cannot persist indefinitely without major corrective or compensating changes. These grim facts coupled with increasing global communication, rapid social and technological change and rising expectations throughout all nations add fuel to the fire of social and political instability. As noted by Eric Hoffer in The Ordeal of Change:

"...... a population subject to drastic change is a population of misfits-unbalanced, explosive, and hungry for action. In other words, drastic change under certain conditions, creates a proclivity of fanatical attitudes, united action, and spectacular manifestation of fluting and defiance. It creates on atmosphere of revolution. The next decade may decide the future of the earth as hesitation for humans. Already our impact has been sufficient to degrade the soil of millions of hectares to thin the ozone layer threatening the rain-forests and to initiate global warming the full consequences of which cannot yet be calculated. At some point in the not too distant future the changes in the environment may cross the threshold of catastrophe. Again we see the intimate relationships between ecology and human affairs. It does not imply that ecology can provide the answers to all these problems but surely a failure to recognize their ecology background invites certain disaster.

Indian's Population Situation

India's population, as on March 1, 1992 stood at 843.93 million (437.60 males and 406.33 million females). The second largest population country, India is now the home of 16 percent of world's population. With this, every sixth person in the world is now an Indian. The country however accounts for 2.42 percent of the total world area. Population counts in India in the modern sense started in 1881 since then, a population census has been conducted regularly every ten years. A census of population was conducted during 1871-72 but it was nosynchrononus. Moreover, the 1972 census omitted served territories, the population of which totaled some 33 million in 1881. The population of India as recorded at each decade census form 1901 has grown steadily except for a decreased during the decade 1911-21. In absolute terms the country population has increased by 160.60 million from 680.3 million in 1981 to 843.9 million in 1991. This is ten times the population of Australia and is more than the population of Japan.

An encouraging feature revealed by the census result is the decline in the growth rate of the population in the last decade compared to 1971 to 1981. It marginally decreased from 24.66 percent recorded in 1971-81 to 23.50 per cent during 1981-91. In most state the growth rate declined during the decade. However, Andhra Pradesh, Arunachal Pradesh, Madhya Pradesh, Maharashtra, Nagaland, Jaipur, West Bengal, Daman and Diu, Lakshadweep and Pondicherry which account for one third of the country's population recorded increase in growth rate. Nagaland registered the highest growth rate of the 56.86 percent while Kerala registered the lowest 13.98 percent. Uttar Pradesh continues to be largest state population wise with 16.44 percent people of the country living there (138,760,417), followed by Bihar with 10.32 percent (86,338,853) and Maharashtra with 9.33 percent (78,706,719). West Bengal occupied fourth place accounting for 80.06 percent (67,982,732) of the country's population followed by Andhra Pradesh with 7.86 percent (66,304,854) Bombay continued to be most populated city of the country with urban population of 12.57 million, Calcutta taking the record place having 10.86 million, Delhi ranks third with a population (urban) of 8.38 million followed by Madras 5.36 million.

Population Density

One of the important indices of Population concentration is the density of population. It is defined as the number of person per square kilometer. The population density has gone up from 216 in 1891, to 267 person in 1991. In 1901 it was 77 person. Density is the highest in Delhi (6.319) followed by Chandigarh (5,620). Lakshadweep which has the smallest population size is the third most densely populated area with a density of 1,615 person per square kilometer, Pondicherry ranks fourth (1605) followed by Daman and Diu (906) Arunachal is the most sparsely populated area with a density of 1,615 person per sq kilometer. The ten heavily populated districts of the country are Calcutta Madras, Greater Bombay, Hyderabad, Delhi, Chandigarh, Maoe, Haora, Kanpur City and Banglore. All of them have density of above 2,00 person per sq km and 5.0 percent of the country's population lives in these districts. The average density of these 10 districts. The average density of these 10 districts is 5,791.

Sex Ratio

In any study of the population, analysis of the sex composition plays a vital role. The sex composition of the population is affected by the differentials in the mentality conditions of males and females, sex relative migration and the sex ratio at birth. The sex ratio is defined as the number of females per 1000 males. In India it has generally been adverse to women. The ratio has also declined over the year except in 1981 when it slightly improved to 934 from 930. In 1991, there has been a fall by five points to 929 per 1000 males. But Kerala represented a different spectrum. The State has a higher number of females than males, 1040 females for 1000 males. In Indian context, a sex ratio of 950 or above can be considered as favorable to females.

Demographic Transition in India

Many scholars have inferred from the actual experience of several European countries and the United States of America that populations historically pass through different more or less well-defined stages in terms of their levels of fertility and mortality. As a generalized explanation of process of mortality and fertility decline in those countries, the transition theory refers to a stage of high fertility in those countries, the transition theory refers to a stage of high fertility and high mortality in simple, mostly agarian, societies which finally transforms itself to a condition of low fertility and low mortality after passing through several other stages.

Today the transition theory it sought to be equally applicable to the less developed counties which are still in the early stages of demographic change. In effect this theory has come to be increasingly regarded as theory that might anticipate the future demographic trends in countries which are currently having high population growth due to rapidly decreasing death rate and high but more or less stable or only slowly declining birth rate. Although there are several expositions of demographic transition theory, but the one dividing the total period into five stage is more commonly used. These stages are as below;

(a) The high stationary stage characterised by high birth rates and high death rates;

(b) The early expanding stage with high birth rates but declining death rates;

(c) The late expanding stage with falling death rates but more rapidly falling birth rates;

(d) The declining stage with low mortality now natality and deaths exceeding births.

The first stage which is characterised by agrarian low income economy is marked by relatively stable birth rate at the maximum fertility level found in societies not using contracepitives, and the death rate is found to fluctuate in response to conditions of prosperity or famines, epidemics and other disasters. Then, as the economy progresses to become more interdependent specialised and market dominated with the generation of agricultural surplus, the average death rate begins to decline under the impact of better organisation and improved medical knowledge and care; however, the birth rate continues to remain almost stable at the earlier level. This leads to an increasing growth rate of population which is the early expanding stage.

As the death rate continues to decline with further improvements in living conditions birth rate also starts declining first slowly and somewhat later at a faster speed making the 'late expanding stage'. In this stage the growth rate starts getting smaller and smaller than what it was in the early expanding stage. In the fourth stage when further reduction in death rate becomes harder to attain, the death rate stabilizes at a low level(between 8-12 per 1000 persons), the birth rate as approaches equality with the death rate and a more gradual growth rate is established when both birth and death rates are low but birth rate fluctuates some what from year to year in response to voluntary decisions of the people. In the last stage of demographic transition which started to appear in recent years in a few European countries like Germany, Sweden, Switzerland etc., the birth rates has declined even below the replacement level and there may not be enough people in the next generation to replace the present one. For India, we now have a history of around 100 years of changes in birth and death rates. The first stage of demographic transition in India continued till about 1920 when both death rates and birth rates were high. This stage was also marked by purely agrarian economy with hardly any significant industrialisation. High mortality levels were hardly and for the variable growth rates of the population during the 30 or more years before 1921. A severe famine affected large areas of the country in 1896 and 1897. In the Bombay Presidency (excluding the princely states) especially, the effects of the famine were aggravated by a severe plague. As a result of the two scourges, the 1901 census recorded a population loss of about 2 percent from the 1891 level of 18.8 million.

Like the 1891-1901 decade the next decade also witnessed several local famines and a severe one in 1907 in most parts of Uttar Pradesh. Plague was in evidence in the Bengal and Bombay Presidencies, and both plague and malaria were widespread in the Punjab and Uttar Pradesh where population growth was negligible. Yet, because the country as a whole suffered from famines less widely and for shorter durations, there was an appreciable rise in the population growth rate compared to the previous decade. During the 1911-20 decade India suffered from an influenza epidemic that caused an estimated 7 percent of the total population to die and led to an absolute decline in the population of the Indian Republic (according to the present boundaries) by 0.77 million at the 1921 census. The second phase of demographic transition in the country started with the beginning of the 1920s and has spread well up to 1971. This is also the period when industrialization began in the country and took some roots. During this period the major causes of high mortality-famines and epidemics-have been gradually brought under control and between 1921 and 1951, the country witnessed a gradual rise in population growth rates. The decline in death rate became sharper during the 1950s without any appreciable decline in birth rate resulting in a population growth rate of 1.96 percent per annum during 1951-61 in comparison to 1.25 percent during 1941-51 decade. During1961-71 the birth rate still remained quite high and due to further decline in death rate the growth rate increased to 2.05 percent. India seems to have entered the third stage of demographic transition since 1971 when birth rates has also started declining. During the birth rate leading to a plateau in the population growth rate during the 1960s and the 1970s. In the 1980s, probably the decline in birth rates was slightly greater than that in mortality; consequently the growth rate during this period is slightly lower than in the previous decade.

Urbanization

Only one in ten people lived in cities when the century began more than half will by the century's end. Today the vast bulk of urban population growth occurs in developing countries. The populations of Third World cities, now doubling every 10-15 years, overwhelm government's attempts to provide clean water, sewage, adequate transport and other basic services. Cities in Industrial countries are gaining people much more slowly, yet even as their growth rates slow or core population decline, these urban areas continue to spread out ward. Industrial and developing countries alike, chaotic, uncontrolled urban growth-whether measured is risking numbers or in the amount of space humans spread out upon—draws on ever more land, water and energy from surrounding regions to meet people's needs. The way cities physically evolve—and the way their development is planned has profound impact on human either recognize the limits of the natural environment or it can destroy the resources on which current and future societies depend, it can meet people's needs equitably or it can enrich some while impoverishing or endangering others. The world needs an urban planning ethic that is sensitive to these environmental and human dimensions. According to 1981 census, India's urban population was 160 million ranking fourth among the countries in the world in terms of absolute size of the urban population, the first three being China, U.S.S.R. and U.S.A.

Though the level of urbanization measured as the percentage of urban to total population has increased from 11 percent in 1901 to only 23 percent in 1981 the number of people living in urban areas and multiplied six folds during the period from 26 million, constituting more than one third of the country's population of 926 million at that time. The growing chaos in urban areas are due to rapid immigration rather than natural increase. A total of 15.7 million persons have migrated from rural to urban areas in the country during the decade 1971-81. The immigrants in the reverse direction i.e. urban to rural numbered. 6.4 million giving rise to a net addition of 9.3 million people in urban areas. There are 218 class I cities in India with more than a lakh population each and these account for about 60 per cent of urban population. Among these, there are 12 metropolitan cities viz., Calcutta (9.16m), Greater Bombay (8.22m), Delhi 5.7m), Madras (4.27), Bangalore (2.9m), Kanpur (1.68m), Pune (1.68m), Nagpur (1.29m), Lucknow (1.0m) and Jaipur (1.0m). With the data on class I cities and the emergence of new towns on the eve of the 1991 census, we can examine the likely level of urbanization in India. It is well known that the share of Class I urban population has constantly agglomerations and cities in the total urban population has constantly increased over time. If one looks at the total urban population has constantly increased over time. If one looks at the trend in India (excluding Assam) from 1951 to 1981, this share increased from 44.7 per cent to 60.8 per cent, the increase being 6.8 per cent during the 1970s. The increase in the share was higher during the 1950s since a very large a number of small towns (857 in all) were declassified on the eve of the 1961 census due to introduction of a restrictive definition of urban areas. In contrast, there was a net addition of 895 towns (or 782 urban agglomerations and towns) between 1971 and 1981 which involved reclassification of almost 250 towns declassified on the eve of the 1961 census. Addition of a large number of small and medium towns during the 1980s brought down the pace of increase in the share of population of Class I cities in 1981.

As the number of new towns indentified on the eve of the 1991 census is quite large, one may expect a substantial increase in the population of medium and small towns in 1991. If, however, a large number of them appear as part of urban agglomerations of Class I cities, their population will be counted as population of Class I cities. An examination of the list of new towns and of cities and urban agglomerations indicate that a large number of new urban settlements have appeared as part of urban agglomerations.

Accordingly, we may expect the share of the population at 65 per cent. This is in view of the fact that 80 Class I cities have been added during the 1980s. We may also not forget that the decadal growth rate of Class I cities slightly declined over the decade compared to the previous decade. Assuming that the proportion of population of Class I cities in the total urban population of 1991 is 65 per cent, we get the total urban population as 213 million excluding that the Assam and Jammu and Kashmir. If we add the 1991 population of Class I cities of Assam, the estimate of total urban population would remain at the 1981 level (that is, 60.5 per cent), the 1991 urban population (excluding Jammu & Kashmir) works out to 230 million, almost the same as projected by the Expert Committee on Population projections in 1984. Thus, the urbanization rate in 1991 varies between 25.6 per cent to 27.5 per cent. Keeping in view that the number of metropolises has increased from 12 in 1981 to 23 in 1991, and that the number of Class I cities has increased from 216 to 296 during the same period, one may expect a higher share of Class I cities in the total urban population than 65 per cent. If this is found to hold good when the data on rural and urban population are released, the total urban population would be less than 214 million.

Population in the twentieth century has exploded particularly in the developing countries. Infact, before 1800, no country was predominantly urban. But by the year 2000, over 50 per cent of the world's population will probably live in urban places and there will probably be more than 250 cities of over one million inhabitants. There has been a world wide trend towards urbanization, the process characterized by the movement of people from rural to urban settlements, from small towns to large cities, and from large cities to the suburbs. This immigration is caused by and in turn leads to profound societies change.

Problem of Housing

The Indian cities are growing both vertically (multi-storey building culture) and horizontally (invasion of the peripheries). In spite of cities expansion there is growing shortage of house (3.8m), growth of slums and squatter settlements, creating interpersonal and intercity imbalances, breaking down services and acceeutuation of the problem of urban renewal. Slums are now becoming permanent features and are proliferating and sheltering larger population every year. They are having small tenements, narrow lanes, lack of sanitation, water supply, electricity, unhygienic environment, etc. Due to realization of the futility of spending money on their clearance, the Government had decided to accord them recognition and gave their residents whatever amenities the funds allowed. Bombay is estimated to have the largest slum population of 3.3 m followed by Calcutta 3.2m, Delhi 2.6m, and Madras 1.8m.

With an unabated inflow of job seekers, these cities are expected to see continuing rise in their slum population. By 1990 when Calcutta will have a population of 12.5m and its slum will be harboring 4.4m Madras population will be 11.8m, 9.7m and 6.0m and of them 4.1, 3.1 and 2.1m slum dwellers. Amongst States, Uttar Pradesh will have the largest slum population of 6.5m followed by Maharashtra 6.2m, West Bengal 4.9m, Tamil Nadu 4.2m, by the end of the decade. The Government's initial strategy was to clear the slums and rehabilitate their dwellers. This had to be given up by the end of the sixth plan as there was constant shortage of funds and developed land. The emphasis was shifted to an on going central scheme known as "Environmental Improvement of Urban Slums," envisaging a package of services for the immediate removal of unhygienic environment of the slums. The services included drinking water, community bathrooms, street-lighting, sewerage, storm water drains and street paving.

Traffic Problems

The traffic problem in urban areas is one of the greatest problems of the day. Most of the cities and towns are growing in unplanned way. Due to lack of proper enforcement agencies encroachment in all parts of the cities is going rapidly and due to political interference, authorities become in competement. Most of the roads are narrow, haphazard and ill-maintained. Traffic problem is getting bad to worse because of high density of population desire to have personal means of transport and to live near the city, mixed traffic, lack of civic sense, ill-maintained road, inadequate expansion in the existing road network poor regulation of traffic, etc. Delhi had 12,000 vehicles in 1951 and the number had shot upto 9 lakh in 1986 and this is expected to reach 40 lakh by 2001. Apart from the fast moving vehicular traffic, the problem on the city roads is being further aggravated because of 50,000 or more slow moving vehicles such as tongas, rickshaws and bullock carts and 20 lakh cycles. There is also a growing trend of road accidents. There were 6,037 reported road accidents of which 1,162 persons lost their lives in 1986.

Shortage of Open Space

Due to increasing commercial value of the land in cities, open spaces are diminishing, even though green spaces are the lungs of the cities, the guarantors of a healthy and aesthetic environment. In Japan, urban parks are classified under two main categories; parks with structures or city parks and zoned green spaces, which are further subdivided into green space conservation zones, agricultural green space and special historical land space zones/scenic beauty conservation zones. City park land includes playgrounds, open spaces, planting strips along side road and other lands awaiting full development. It is commonly said that these parks become the pursuit of undesirable activities including sex, crime, gambling, etc. In spite of these negative aspects, it provides thousand times much needed benefits for the city dwellers. There is a concept to have at least 20 per cent area of the cities under open space in the form of green belts, parks, playgrounds, etc. But it has become an utopian concept. In 1973, master plan of Allahabad was approved and much of the areas were put under green belts. But due to illegal spraw of colonies most of that very areas were converted into concrete belts. The green belts concepts are also not getting proper place in the city environment improvement scheme because of lack of financial support, public interest and professionally qualified staff. Besides this, there is not definite framework for parks and recreational areas in relation to national goals. City parks and green belts are developed mostly by Municipalities or Corporations, Social Forestry and Horticulture departments. Private organizations, voluntary organizations, industries should also come forward to set city parks and green belts. Vacant lands acquired by the Corporation are to be sued not only for commercial or residential settlement but also for establishing big parks.

Local communities must come forward to participate in the management of parks and can also raise funds for it. Country parks can be established in rural areas to provide a better place for recreation and enjoyment of rural people. In addition to parks and gardens, landscaping measures should be applied on flyovers, overhead bridges, wall and other structures, such as the growing of creepers, climbers, scrubs and transplanted trees. There is a need for making legislation for City and Country Parks Act for the forceful implementation of the concept of green belt.

Urban Planning Around the Globe

Virtually all cities share some land related concerns, such as congestion and pollution from motor vehicles, lack of affordable housing, and the cancerous growth of blighted district. Of course, prescriptions for better urban planning are not the same for all parts of the world. Many land use issues in industrial countries seem mainly concerned with the quality of urban life, and those in Third World cities are often questions of life and death. But all cities, whether surrounded by affluent suburbs or makeshift shanty towns, now need to plan land use far more carefully than in the past before the developing world's urban crises than into catastrophes, and the industrial world' urban crises turn into catastrophes, and the industrial world's problems become issues of survival. The extent to which cities control the use of land varies widely. Japan and Western Europe have the world's most comprehensive urban land use controls; their cities are generally compact, laid out for efficiency in transport, water and energy use, and building materials. Japan protects farmland with strict zoning and tax policies. Its suburbs are of moderate, not low, density.

Much of Japan's urban activity is concentrated in Tokyo and along the dense Tokadio corridor between that city and Osaka. This compact, linear form has allowed great efficiency in transport, including development of the world's first high-speed train, which has been running since 1964. However, the country's 33 million automobile—nearly twice as many as in all other countries combined—still plague Japanese cities with extreme traffic congestion, especially in the Tokyo area. Western Europe has a long tradition of activity controlling land use so that the small amount of available space serves the public's interest more than that the private developers. Paris, a city for nearly 2,000 years, has planned and regulated land use since the Middle Ages. Many of England's urban areas, still fulfilling a far sighted decree by Queen Elizabeth I in 1580, are ringed by green belts intended to protect farmland and prevent sprawl. But in post war decades the automobile has had a profound effect on Europe's compact character, fuelling suburbanization.

In the fifties and sixties, some older European cities embraced the U.S. model, demolishing corner shops and cafes to make way for urban freeways. Later, planners tried to contain suburbanization by diverting growth into smaller satellite towns. Since the mid-seventies, they have turned their attention to revitalizing shrinking inner cities. Control of land use in Eastern Europe and the Soviet Union has not been as effective as might be assumed for centrally planned economies. Urban planning therefore is not well coordinated with economic decisions, and sometimes reflects rivalries among government ministries.

Extreme water air, and soil contamination from poor placement and regulation of polluting industries is a critical concern. Although urban planners in these countries historically have supported public transport and promoted urban densities high enough to make it viable, in recent years and use has become less efficient. Nearly all new housing has sprung up at the edges of cities, with few accompanying shops, jobs, or schools, resulting in excessively long commutes. On going political reform and the region's cities will become more automobile-oriented, perpetuating the environmentally destructive patterns of western industrial cities. Among industrial regions, North America and Australia have the weakest planning traditions. Governments on these continents have done relatively little to guide development beyond separating industrial areas from those zoned for commerce and housing.

Many have enforced low-density zoning in an effort to curb urban expansion—a move that, ironically, achieves just the opposite. Zoning codes that restrict residential density, usually by requiring each house to occupy its own large lot, have forced development to consume even greater tracts of open space. And to accommodate new, outlying communities, roads have been extended further outward, attracting even more spread out growth. Even older U.S. cities are becoming more dispersed now than in the pre-automobile area. Growth in the New York metropolitan region has turned to sprawl; while population has grown only 5 per cent in the past 25 years, the developed areas has increased by 61 per cent—consuming nearly quarter of the region's open space, forest, and farm land. Many people move out to the suburbs and exurbs seeking open space and bounds with nature that come only in a rural setting. Yet most of these residents continue to maintain urban lifestyle—communiting to jobs in the city and demanding urban amenities from a suburban shopping mall near their home. The result is neither urban nor rural living but a destructive compromise that the environment sustain. Developing countries have the loosest controls over how cities develop. Local governments in Asia, Africa, and Latin America often have neither the authority to guide land use nor the funds to provide basic services. Municipalities are fiscally dependent on central government, which are themselves struggling under crushing debt burdens. With few exceptions, urban planning is a relatively recent phenomenon in developing countries, even in the capitals.

In India it reportedly began with Delhi's Master Plan in 1962. In Nigeria, the first systematic effort to plan cities did not begin until 1971; the capital, Lagos, still has no designated metropolitan government and no formally defined urban area. The Lagos State Ministries that have their own priorities. The Third World is burdened by several enormous, rapidly growing cities—including-Sao Paulo, Shanghai, and Mexico City—whose sheer size and instability create problems on an entirely different scale. As early as 1964, researcher Ronald Wraith used a single word to denote these giant cities racked with pollution and rimmed by shanty-towns: "Megalopolis-the city running riot with no one able to control it." Since then, megacities have become increasingly characteristic of the developing world. In 1950, only three of the world's 10 largest cities were in the Third World by 1980, seven of them were there.

Although skilled land use planning is badly needed in these megacities, its effectiveness is limited in the absence of other kinds of change. One reason is that most of the physical growth takes place in illegal unplanned squatter settlements, rendering useless even existing mechanisms for guiding land use. Called fevelas in Brazil bidonvilles in French-speaking West Africa, Spanish in the Middle East, and kampungs in Indonesia, these illegal communities hold 30-60 per cent of the population of many Third World cities. The unhealthy conditions in these settlements can only be addressed fully through extensive economic and social reforms that attack the root causes of poverty not just in cities but also in the rural areas that urban migrants abandon in search of economic opportunity.

The effectiveness of land use planning is also limited in the absence of family planning programs. (Only in Africa does migration account for the increase in much of Asia and Latin America results from the fertility of people already in cities). Finally, planners in the Third World's giant cities faces colossal environmental problems, from deadly air pollution in Mexico City, to indiscriminate dumping of toxic wastes in Alexandria, to the actual sinking of cities such as Bangkok, Jakarta, and Shanghai due to over-drawing of ground water.

Crowding

Crowding is a conspicuous ecologic force impinging upon modern man. It results from more than a simple increase in population density. It is also a product of activity patterns psychological needs, and social contact. Crowding implies a force, a pressure and psychological reaction. No clear definition of crowding is possible at the present time; it remains an operational concept with many subjective qualities. Crowded population are, however, easier to recognize than to define, and they are studied more carefully, better definitions may become possible. For the time being, we might define crowded populations as those in which there seem to be excessive number of individuals per unit space in relation of the activity of the industrials and the quality of the environment. With an operational concept of this type, ecological studies have shown that some rodent populations may be "crowded" in relation to their social behavior and habitual requirement at a density of 5 individuals per acre, where as other at the same species may not be "crowded" until they have a density of several hundred per acre. Species differ in this regard, and so also individuals and social groups within the same species.

A western cattle rancher may feel "crowded" at a density of 10 homesteads per square mile whereas a modern suburbanite may feel relatively "uncrowded" at a density of several hundred homes per square mile. The western rancher ranges several square miles per day over what he considers his personal property and he does not welcome potential competitors in his territory. Conversely, suburbanite restricts his daily movements, apart from going to work and to shop, to a lawn around his home, and he does not view neighbors as a competitive threat.

Hence, in both animal ecology and human sociology, it is important to distinguish between population density per se, and crowding, since the latter depends so much on behavioral factors. The spatial requirements of man are obviously very variable, and result from both individual difference and social conditioning. It is impossible to state what the optimal spatial and social needs of man are as those of the Japanese city dweller ar much different than those of the Kansas farmer. Nevertheless, it is safe to assert that man is now subjecting himself to increased crowding through the combined forces of population growth, urbanization, transportation and mass communication. Man's perception of crowding has skyrocketed upwards in recent years. The growth of urban traffic and the increased number of people in public buildings, schools, hospitals, stores, exhibitions, recreational areas, etc., have all contributed to the feeling of population pressure. Through the mass media or audios, television and newspapers, individuals are vastly more aware of worldwide events. Social pressures at any point of the globe are quickly felt by all nations.

Local crime, violence, and warfare now become worldwide worries in a matter of hours. The people of one continent feel threatened by those of another continent thousands of miles away. The inevitable result has been a great magnification of social anxiety and population pressure. The world has thus become more limited in its psychological space, and man more crowded in a social and behavioural sense. How does one evaluate crowding as an ecologic force? It definitely requires knowledge of at least three major parameters: (1) population density, (2) environmental structure, and (3) social behaviour in relation to each of the possible in both animals and man, but in both, especially man, it is often possible to isolate crowding as the key variable. Many other factor are interwoven.

Crowded human populations are often so characterized by poverty, malnutrition, lack of educational and recreational opportunities, varies environmental hazards and unstable social patterns that it is virtually impossible to separate crowding itself as the one key variable. The essential question becomes, how does one measure and analyze crowding per se apart from all it related phenomena? It may also be important to distinguish between crowding of various origins. Since man often seeks social contact and stimulation, does not crowding contain constructive and creative social elements, or is it inevitably a factor in increased social problems and environmental deterioration?

From man himself the answer to these questions are confusing and uncertain. The behavioral sciences of man have just begun to investigate human spatial behavior. More direct clues have come from animal studies where experimental work has been possible. The simpler social systems of many animals have permitted more direct observation of behavior, more ready identification of variable, and more penetrating analyses of causes and effect relationships.

In animal studies it has been possible to distinguish between extrinsic crowding; that crowding imposed on the animals as an external force, as in captivity, compared to intrinsic crowding in which dense aggregations result from the social behavior of the animals. A similar distinction is sometimes possible in human populations, but frequently it is difficult to discern if people are crowded by choice or the external pressures of economics and society. It was demonstrated more than 30 years ago by Professor W.C. Alee and his students of the University of Chicago that most species of animal have optimal levels of crowding, above which and below which deleterious effects occur. These studies showed that certain levels of crowding are necessary to maintain normal biologic function and social activity.

Goldfish, for example, grow most rapidly and live longest in social groups of certain densities compared to isolated fish to fish in smaller groups. Fresh water shrimp survive longest and tolerate harmful environmental conditions most successfully at certain optimal stage of crowding. Similarly bobwhite quail survive harsh winters best and reproduce most successfully in the spring in conveys of certain optimal size.

Beyond optimal levels of crowding, however, benefits are quickly lost and various forms of pathology, occur. Growth may be stunted, life span shortened and various diseases or abnormal behaviors appear. Most studies on crowded animal populations in the last 20 years have concentrated primarily on the harmful effects of overcrowding. Such effects usually involve both physiologic and behavioral abnormalities. The physiologic effect of overcrowding centre around endocrine imbalances of the adrenal pituitary system. These imbalances can lead to a great variety of disease condition including gastric ulcers, hypertension, nephrosclerosis, arteriosclerosis, and increased susceptibility of infectious diseases.

Prolonged exposure to crowding and its associated stresses have produced additions disease (hypoadrenalism or adrenal exhaustion) myasthenia (muscular weakness), and a variety of serious metabolic disorders leading to shock, coma, and even death. It should be pointed out that these stress effects are quite variable, however, and do not consistently reflected the same patterns in different individuals and various social groups. Behavioral pathologies also appear in experimental animals as a result of crowding and these are sometimes manifested more quickly than are the physiologic changes. Overcrowded animals often show increased aggression and violence, abnormal sexual behavior, disruption of normal nest building and maintenance, breakup of stable social groups, disappearance of alteration of normal social roles, parental desertion of young and frequently cannibalism of young. The leading question, of course, is whether or not similar processes occur in man as a result of overcrowding. There are so many variables in man which affect health and behavior, that this becomes a very difficult question. There is no doubt that some striking correlations of disease and behavior are associated with urban concentrations. Many human problems seem to be so accentuated in crowded populations that they virtually constitute an "inner city syndrome" disease, and emphysema; maternal problems such as high prenatal loss, parturitional difficulties and high infant mortlity; behaviour problems such as alcoholism, drug abuse, mental illness, and criminal assault; and social problems such as high divorce rates and social instability.

More recently it has been found that high prevalence of mental illness in the lower socioeconomic classes and dense urban ghettos of several other cities. The three volume report on Manhattan entitled, "Mental Health in Metropolis," also found unusually high prevalence of mental illness in crowded inner city inhabitants. The problem of urban man seem so strikingly similar to those effects produced in overcrowded animal populations, it is tempting to conclude that similar processes are occurring. But, as stated previously, many factors are interacting with crowding in urban manpower, malnutrition, lack of educational and recreational opportunities, various environmental hazards, and unstable social patterns—that it is virtually impossible to separate cause and effect relationships. There are enough solid exceptions to the principle of "high density produces high pathology" that one must be cautious in carrying this dogma too far. For example, the young nation of Singapore is obviously crowded—over 2,000,000 people on an island of 224 square miles, making it one of the most densely populated nations of the world—yet it enjoys remarkable good health. The death rate is one of the lowest in the world (5.5 per 1000) and the prenatal mortality is lower than that of Britain and the United States.

Major factors in this good record many be its youthful population (40 per cent of the Singapore population is under 15 years of age), and the well planned interspersion of parks and greenbelts with residential and commercial areas. One does not feel a "crowded" in generalizing as in Hong Kong, Tokyo, New York and Calcutta. Such variable demonstrate the difficulty of generalizing broadly about crowding or population density as a pathologic influence. Thus, the precise role of crowding in the "inner city syndrome" must remain a cloudily issue until more and better research studies are available.

An encouraging development of recent years is the awareness on the part of some architects, city planner anthropologist and others of the relationship between our environment and mental health. In Architectural Environment and our Mental health, Moiler explores the impact of physical space and structure on the physiological well-being of man. He noted that more buildings will be built in the next ten years than have accumulated since the beginning of civilization." And he wisely emphasized that, "We should recognize that if we continue to permit the structuring of a physical environment which is essentially hostile, to which man must somehow try to adopt, the cost to individuals and to society will be far too grat—in terms of mental illness, delinquency, poor motivation and fulfillment of our capabilities for creative work and community usefulness."

Anthropologists have also been interested in the spatial behaviour of man, and the book by Professor E.T. Hall entitled "The Hiddeen Dimension," was the first one to highlight some of the behavior patterns of man relating to social spacing. Hall showed the existence of "personal space" individuals and pointed out cultural differences in the extent and importance of personal space. A book edited by Esser reviewed the interest of psychiatrists, sociologists, biologists and psychologists, in the relationship of ecology and environment of human behavior. It emphasized how little we know about ourselves in these vital matters. Although most of the data on human spatial behavior are preliminary, it is a hopeful sign that several professional and scientific disciplines are now giving attention to this topic. We can be assured that the ecology of crowding will become increasingly vital to man's health and welfare. The immense capacity of populations for increase cannot be better illustrated than by that of the human population, which has grown prolifically, at times doubling each quarter-century. From the earliest time that the rapid increase in the number of humans could be appreciated, population growth has caused concern.

The English economist Thomas Malthus, in his essay on population published in 1798, warned of the dangers of unchecked population growth. He began, "I think I may fairly make two postulata. First, that food is necessary to the existence of man. Secondly, that the passion between the sexes is necessary and will remain nearly in its present state." He went on to note that "Population, when unchecked, increases in a geometrical ratio: Subsistence increases only in an arithmetical ratio. A slight acquaintance with numbers will shew the immensity of the first power in comparison of the second." By contrasting geometric growth and arithmetic growth, Malthus demonstrated an appreciation of the most basic property of population growth; exponential increase has been the foundation of all efforts to derive formulae for the projection of population size through time.

A number that increases arithmetically is added to by a constant amount during each interval of time. When a worker shovels coal into a bunker, the coal therein increases at a constant rate, say 1 ton per hour, regardless of the amount accumulated. That is, the growth of the coal pile is determined, not by the amount either already shoveled or still remaining, but by the pace of the worker. Accumulation rate is constant. Growth is arithmetic. In contrast, geometric (or exponential) increase occurs in proportion to the amount accumulated. The larger the population the more rapidly it grows, because increase in numbers depends upon reproduction by individuals in the population. For coal, this would require some attraction of the chunks in the bunker for those remaining in the pile outside.

For populations of plants or animals, Malthus's "passion between the sexes" is quite sufficient. Greatly impressed by Malthus's arguments on population. Charles Darwin wrote, in On the Origin of Species, "There is no exception to the rule that every organic being naturally increase at so high a rate, that, if not destroyed, the earth would soon be covered by the progeny of a single pair." To make his case as forcefully as possible, Darwin put forth a conservative example:

> The elephant is reckoned the slowest breeder of all known animals, and I have taken some pains to estimate its probable minimum rate of natural increase; it will be safest to assume that it begins breeding when thirty years old, and goes on breeding till ninety years old, bringing forth six young in the interval, and surviving till one hundred years old; if this be so, after a period of from 740 to 750 years there would be nearly nineteen million elephants alive, descended from the first pair.

Because baby elephants grow up, mature, and themselves have babies, the elephant population grows exponentially. Imagine, however, a population in which only one female reproduces; such a population, like the coal pile, would increase arithmetically, and very slowly at that as "a slight acquaintance with numbers will shew."

6 Population Ecology

One of the earliest terminological distinctions made in the emergence of "self-conscious" ecology was the addition of prefixes to create "autecology" and "synecology" Aut-literally designates "self" or "individual," but in ecological practice it described studies of a small group of individuals of a species, which was regarded as unit. Ambiguity was avoided, temporarily, by Adams who distinguished three classes of ecology: "individual. " aggregate, "specifying the taxonomic unit, and "associational," citing Mobius's biocoenose as an example of the last. Adams's contemporary, Shelford had little use for autecology.

To Shelford, "Ecology is a science of communities. A study of the relations of a single species to the environment conceived without reference to communities and, in the end, unrelated to the natural phenomena of its habitat and community associates is not properly included in the field of ecology." Plant ecologists at the Third International Botanical Congress in 1910 formally adopted the terms *autecology* and *synecology*: Ecology, the study of conditions of environment and of adaptation of plant species is if taken isolately "autecology," if taken in association "synecology." Chapman confirmed this meaning of *autecology* for animal ecologists, commenting that it was this aspect of ecology, bordering on physiology, which even raised the question of whether ecology was a distinct science. However, turning to the subject matter of synecology and the interactions of species and the complex of environmental factors, Chapman clearly recognized "a field no longer that of general physiology" but "more distinctly as that of ecology."

Chapman's question about the distinctiveness of ecology at the autecological end persists particularly in a recent hope to merge it with population genetics. The overlapping interests of physiological ecologists in physiological attributes of populations, of population geneticists in genetic variability of populations and their evolution, and of populations are currently included under the term population biology. Self-conscious ecology emerged at the turn of the century, its founders claiming a relationship with natural history and evolution, especially in explaining adaptation, and also linking it to physiology. Notwithstanding Haeckel's definition of ecology as the relation of organisms to their environment, explicitly including other organisms, much protoecology and early ecology concentrated on relations of organisms only with their physical environment, explicitly including other organisms, much protoecology and early ecology concentrated on relations of organisms only with their physical environment.

The zoologist Semper for example, considered the effects on animals on temperature relations of animals. The botanists Warming and Schimper emphasized water relations of plants. Pound and Clements MacMillan's classic contribution to plant ecology, noted, with some disapproval, that he was solely concerned with the physical environment. The earliest physiological-ecological work in America funded in 1903 by a private foundation, the Carnegie Institution, stressed measurement of the physical environment, and physiological responses of plants at the Desert Laboratory and provided a microcosm of ecology emerging from physiology would provide the touch stone for ecology, or even make it unnecessary, so that it would merge with physiology. No merger occurred; physiology and ecology separated, but produced hybrids, or at least "hopeful monsters." Segregating ecology from physiology and, subsequently, distinguishing the several hybrids, physiological-ecology, ecological-physiology, and cephysiology, is not easy Earlier terms were abandoned. "Response physiology" was concerned mainly with behavior of animals; "relations physiology" or toleration physiology" was concerned with the distribution and responses of a population, normally of one species, to individual environmental factors.

A problem of ecologists is deciding what is the appropriate relation of ecology to its parent science and what physiology contributes to ecology per se or to the several hyphenated combinations. Livingston one of the early staff of the Desert Laboratory, was straightforward:

> By physiological ecology is here meant merely the study of factors which determine the occurrence and behaviour of plants growing under uncontrolled conditions.

Shelford described both controlled laboratory and field experiments. Defining the response of animals to environmental factors as physiological ecology. Shelford was one of the early animal ecologists who studied the community and stressed its significance in ecology. However, he also voiced the hope of other contemporary and later ecologists:

> If we knew the physiological life histories of a majority of animals, most other ecological problems would be easy of solution.

Shelford, perhaps anticipating later controlled environmental chambers, phytotrons, and biotrons, allowed that physiological life histories could be worked out in a laboratory with elaborate facilities." However, he centred on the mores" problem, that is, physiological life histories in natural environments together with other organisms in communities. In 1929 shelford published Laboratory and Field Ecology, and his famous vivarium at the University of Illinois became the seat of a distinguished tradition of experimental and physiological ecology of animals. Ecologists were, and are ambivalent and inconsistent in their use of the terms *autecology, environment*, and *physiological ecology*. chapman distinguished physical autecology and biotic autecology. Some ecologists seeking the historical beginning of physiological ecology found it in organisms as related to the physical environment in the work of Reaumur, Priestley, Lavoisier, and Van't Hoff.

Although all ecologist recognise that environment includes other organisms, the tendency to concentrate on physical factors in more up-to-date phraseology, "variables" or "parameters" is difficult to resist. This may be because they are more easily measured Physiological ecology is diverse if not elusive subject. In one sense it is physical autecology and stresses the internal function of individual organisms or parts thereof. In its more typical sense it concentrates on the function of organisms in natural environments, stressing the entire organism.

A recent survey of physiological ecologists asking "What is physiological ecology?" yielded diverse answers. The great, if not unanimous, consensus was that it involved natural conditions and the whole organism. Early studies of physiological responses of organisms concentrated largely on the response to a physical factor of the environment. Since these studies normally examined the response of a group or population of organisms to the factor, they constituted a facet of population ecology.

The classical rationale of such studies was the so-called law of the minimum formulated by the famous 19th-century German agricultural chemist Justus Liebig. Liebig's "law" stated that the rate of any process is limited by the least, or slowest, factor affecting it. Shellord elaborated on the idea of limiting factors" in his "law of toleration," which asserted that a population or species is limited in its distribution or function by a minimum and a maximum level of any factor, with an optimum between these at which the organism grows and reproduces best.

Various branches of early self-conscious ecology examined the effect of different physical factors on the distribution, growth, and reproduction of populations, particularly those of economic or public health concern. This emphasis on the response of organisms and populations to the physical environment was evident in the organization and emphasis of most first-generation reference works and textbooks in ecology. Chapman for example, devoted 146 pages of his pioneer textbook of animal ecology to physical autecology.

Andrewartha and Birch in a history of insect population ecology, started with ecological physiology, an extended review of the physical environment, and the relation of insects to physical variables. In their view, "The concept of the environment is the very essence of the theory of population ecology." Theoretical population ecology, as it developed in the 1920s and 1930s, had strong links with physiology and environment, and some of its earliest proponents were cognizant of these links. G.F. Gause wrote, The study of the correlation between the density of population and environment represents the leading problem of the ecology of populations" Gause expanded the qualitative concept of Shelford's "law" of tolerance to provide a specific mathematical expression to describe the relation, or degree of "ecological amplitude," between a population and an environmental factor: The correlation between the density of population of very different organisms and ecological factors in natural conditions can be expressed by means of a special type of curve, that of Gauss.Gause joined his Russian compatriot Ramensky, among other early ecologists, in using the mean and standard deviation of a normal Gausian curve as a description of the conditions in which a species grows and Gause used the standard deviation to characterize the ecological plasticity" of a species Physiological tolerance ranges of species populations, with respect to one or more factors, were commonly measured in experimental contexts and in nature where they anticipated "niche breadth" curves.

A major problem in experimental physiological studies lay in the assumption that tolerance ranges and optima, determined in experimental contexts, are effective in explaining or predicting the distributions of species in their normal habitat in relation to many variables and their interactions with other species. Autecology sensu stricto, has limits, since it omits such interactions, although it may provide suggestive evidence. A population is commonly defined as a collection of individuals of a species. In human demography, and etymologically, it is clearly one species, Homo sapiens. In general biology, ecology, and statistics, population is less straightforward.

Population was used by biologists and early ecologists in a collective sense to describe all organisms of a group or area, for example, a population of invertebrates, fishes, or birds. Johnstone like many of his predecessors and contemporaries, defined population as the number of individuals per square foot or referred to the total population of the sea, but he also described a population as the numbers of individuals of each kind of organism with its own birth rates and death rates.

Thomas park a notable experimental population ecologist, considered both single-species" and "mixed-species" populations and saw no difficulty in using the word population for both. Park also provided the convenient terms population biology and population ecology to add to population dynamics a term used earlier by Elton. The dual meaning of population persisted in the leading mid-20th-century reference of animal ecology.

A glossary prepared by the Committee on Nomenclature of the Ecological Society of America defined population as "the sum total of the individuals in a given area; sometimes used with the implication that individuals are of the same species. "The looser use of population, which characterized early ecology, gave way, with some qualifications, to the definition of population as the number of individuals of a species as the unit of choice. Aggregates of species populations had long been called communities reviewed the history of the term and the study of populations and allowed that no one, not even students of classical languages, was offended any more by reference to nonhuman "populations."

By this time, the autonomy of population ecology as an important some believed the important aspect of ecology was secure. Population ecology, as it became a more discrete discipline, usually included the study of dynamics of single-species populations, the interactions between two-species populations with very tentative excursions into three species interactions, and sometimes promised to provide a theory and explanation of multispecies communities. Comments on an interest in populations of given kinds of organism and attributes of populations date to antiquity. These have been reviewed by cole, Egerton and Hutchinson. As might be expected, populations of locusts or disease organisms such as plague attracted attention even when not recognized as populations. Specific concerns about population growth of humans and animals of economic interest are scattered through the literature up to the 18th century without leading to formulation of nay general concepts other than balance of nature. Similarly, interactions between populations were matters of practical and speculative concern to many, among them Benjamin Franklin, who considered the effect of predation by birds on insects and A.P. DeCandolle, who recognized that competition might influence geographical distribution of organisms.

The famous and much disputed exposition of human population growth by Malthus in his Essay on the Principle of Population posed the dire consequences of geometric increase of human populations and the tragic means of their limitation, Darwin and Alfred Russell Wallace both derived from Malthus the basic idea of struggle for existence which stimulated their concept of natural selection and the idea that environment limits populations.

Formal mathematical formulations of population growth appeared with greater sophistication and led to intense activity in theoretical population ecology in the 1920s and 1930s and a grand upheaval in the 1950s and thereafter. Darwin, who anticipated many of the basic concepts of ecology, also presaged the attitude of some biologists who were less than enthralled by mathematical formulations of population growth implicit in Darwin's verbal model. He wrote:

> We have better evidence on this subject than mere theoretical calculations, namely the numerous recorded cases of the astonishingly rapid increase of various animals in a state of nature, when circumstances have been favorable to them during two or three following seasons.

In fact, the forerunner of "mere theoretical calculations"

of population growth antedated Darwin's work. In 1838, going beyond Malthus, P.F. Verhulst formulated the famous "logistic" equation of population growth which like Mendel's similarly premature effort, went unnoticed until it was rediscovered in the 1920s. With perhaps some poetic justice the botanist Karl Nageli, who had failed to appreciate Mendel's work, produced a promising start in the quantitative study of plant populations, which was ignored, like the work of Mendel and Verhulst. Park provided a useful outline of the history of "modern population ecology" stemming from four largely separate sources, "human demography" and "biometry" in one dyad and "early ecology" and "general and comparative physiology" in another. He illustrated each of these as having a laboratory population shunt and fusing in modern population ecology.

Physiological ecology or autecology, as we have seen, is not readily dissociated from modern population ecology, and insofar as it focuses on elucidating processes that influence population growth sharp definition may be unnecessary. Park's "early ecology" may be taken to include general natural history and biogeography. One of their concerns was the distribution of populations of various taxa, including species, and that most important population phenomenon the distinction between absence, or zero, and some, at least enough to be encountered and to record, as present. This distinction was not as dichotomous as it seemed, since zero may be a function of sampling intensity.

The problems of spatial limits, range boundaries biogeography, or local habitat limits in the ecological scale, and the factor accounting for these, were among the earliest and most specific concerns regarding populations and still persists. The distinction of boundaries in time in which a population entered a locality by migration or introduction, or ceased to exist there by migration, extirpation, or, with greater finality, extinction, was another aspect of population studies.

Population census and survey

Park included, in his history of population ecology, quantitative "census" of natural populations as a link between "early ecology" and "modern population ecology." Most early census studies were of many species based on the habitat or the collecting technique and in the modern sense would be community studies. Park recognized that quantitative knowledge of populations was required for community analysis. He also noted that Mobius's community definition "focussed attention on the ineluctable interdependence between the natural populations comprising the community instead of on particular habitat relations exhibited by species. "The problems and interest of population ecology may, in fact, be seen as beginning and ending in community ecology. Slobodkin began his book on population ecology as follows:

> If you were to make a fence around any region of the earth's surface and list the kinds and numbers of organisms found within that fence, you would be starting to define the problem of population ecology.

Ecologists started in just this way, and the population as a group of organisms, of one or several species, with particular properties, and in a specific area was of paramount interest Clearly a population had:

1. Spatial extent, areal and habitat boundaries, and physiological tolerances.
2. Number, variously known as abundance or density.
3. Distribution or dispersion within the area, commonly and conveniently described as uniform or homogeneous but subsequently recognized with much frustration by empiricist and theorist, as heterogeneous or patterned.

4. Individual or collective size, such as area covered or volume or weight per unit area.

5. Jess readily quantifiable attributes, such as vigor or success in completing a population's growth and life cycle, and chronological or seasonal attributes, known as phenology or periodicity.

6. "Aggregation" or social properties among its members, which were extensively studied in animals by Allee.

Allee demonstrated what came to be called the "Allee effect," in which reproductive rate decreased when the population dropped below a certain critical density, contrary to the assumption of the classical logistic model of population growth.

7. "Associational" properties shown by tendencies to co-occur and, presumably, to interact with other populations These, of course, can only be discerned by examining mutual distributions of populations and their interactions in nature, and such studies range from examining two-species relationships, such as pollination, predation, or competition, to multispecies communities.

8. "Dynamic" properties, those that changed as consequence of reproduction, mortality, and dispersal in the form of immigration and emigration.

Empirical population ecologists were concerned with all of these; theoretical population ecologists preferred to ignore questions of *dispersal* or migration, and *dispersion* or pattern of distribution in space. Population ecologists sought understanding of numerical change, population growth, fluctuations or oscillations, stability or equilibrium, and decline or extinction. A major and continuing concern, as Park noted, is "dynamics of production," the amount of biomass produced.Under the rubric of "biological survey" or "census," late 19th century and early 10th century biologists charted the distribution of species or aggregates of species of an area, largely at first in qualitative terms, later quantitatively. Census was a misnomer, since it rarely involved a complete count or was not of a population of a single species, but was usually a sample, sometimes extrapolated to a very large total, of several species. Marine biologists and limnologists were well in advance of terrestrial animal ecologists in sampling populations. Hardy a distinguished British marine biologist, wrote:

> We have a better idea of the population density, per unit area of different molluscs on many parts of the sea floor than we have of the snails of our countryside; our knowledge of the numbers of small crustacea per cubic metre of water in may seas of the world is far in advance of such information regarding insect numbers on land.

Hardy, in fact, claimed that the new science of ecology was more advanced in the sea than on land because marine animals were Rondomesticated and governments had supported fisheries research, "which is essentially ecological." An interesting dichotomy appeared among aquatic biologists which continues to the present. Students of benthic organisms and plankton emphasized communities and synecology, whereas those interested in fishes and aquatic mammals concentrated on autecology. Marine biologists and limnologists-captured organisms in the aggregate by seine.

Dredge, trawl, and net. Hensen and for example, counted the numbers of plankton and benthic organisms, respectively, which sometimes were collectively referred to as populations, as well as being comprised of individual populations of taxa, depending on the current capabilities of identifying organisms. haeckel in addition to questioning the applicability of mathematics to population studies, criticized hensen because, he said the enormous labor of counting plankton had potential for "ruin of mind and body." haeckel advocated rmass and weight and chemical analysis" as the proper measure of a population. Hensen had, in fact, measured weight in addition to counts. Hensen even went beyond the usual collection and counting of adults. His plankton studies grew out of his efforts to determine the numbers of fish eggs and larvae, and he attempted to calculate from these the number of female fish that produced eggs. Such estimation of total numbers of adults required determination of sex ratios.

Quantitative surveys of aquatic animal communities and their included populations proliferated, and the variation of numbers of organisms from place to place and time to time became apparent from these surveys. In these studies, the distinction between studies of populations and quantitative description of communities was etymologically and empirically confounded. Johnstone calculated the density of the fish "population" of the North Sea based on an estimate of 9,000 million individuals of six fish species.

In addition to the usual estimation of density as number per-unit area; he calculated the reciprocal, or mean area per fish, as terrestrial botanists sometimes did for plants, Each fish inhabited an average area of bottom of 60 square meters. Johnstone wrote that accumulated observations pointed to the conclusion, now the conventional wisdom of marine ecologists that the density of both vertebrates and invertebrates is greater closer to land than farther out to sea. He also said that quantitative methods were useful for investigating "dynamical equilibrium" in total sea life and the variability in one or more species. Gross changes in fish populations of the Great lakes also became a matter of concern in the 1920s, paralleling the observations of population declines in marine fisheries. Although population problems were increasingly recognised, ecologists had not made substantial progress toward understanding population phenomena in the early decades of the 20th century. Some marine biologists risked running mind and body and adopted hensen's counting techniques, but like him, they sometimes weighed and chemically analyzed their collections. Some objected to counts because of the errors due to "excessive fatigue" and suggested the time was better spent on estimations of more samples.

Early limnologists were also concerned with populations, their numbers, and their distribution. Birge followed swarms of Daphnia hyalina in surface waters, with between 800,000 and 1,492,000 individuals per cubic meter, which, he said, attracted immense numbers of perch. Birge also counted numbers of plankton and applied the term population to individual species and more inclusive taxa up to the total crustacean population.

He followed the increase in numbers of species in the samples and the vertical distribution of several species, finding that all but one species conformed to a "general law" that numbers became maximum and uniformly distributed simultaneously, anticipating a later search for a principle of numbers of species in a community. Birge and Juday subordinated numerical counts to biomass measurements of plankton, and terms such as standing crop and productivity were added to population and community.

Continuing surveys of plankton and benthos included studies of emergence and populations of insects and even population growth of higher plants.Studies of commercial fish and shellfish populations proliferated because of concern about declining populations in the late 19th and early 20th centuries. Age class determination of fishes by "scale ring" counts and, less familiarly, by ear stones, or otoliths, began in 1898 and 1899, respectively, This led to estimations of rates of growth and age composition of populations and a long tradition of population studies of fish. Ring scales, although not annual, paralleled the problem of tree rings in being less useful in the tropics than in temperate or Arctic waters. Perhaps because of emphasis on commercial species, institutional constraints, and limitations of technique, fish stocks were mostly considered as single species.

Population data of many fish and invertebrates were derived from commercial catches, but there were efforts to address basic population attributes. In the last decade of the 19th century, C.G.J. Peterson, in an early mark-recapture study of movement, marked plaice with bone buttons. Shortly after the turn of the century, J. Hjort, using the new technique of aging by scale counts, followed the age distributions of fish populations, notably herring, and illustrated the effect of year classes on fluctuation in abundance of fish. Hjort urged study of birth rate, age class, and migration of fish. Herrick in his monograph Natural History of the American Lobster, concentrated on morphology, habits, and growth but turned briefly to the need to restore and maintain population equilibrium.

The traditional mechanisms, closed seasons, size limits, protection of females, and release of young or adults were all cited; and Herrick estimated number of eggs per female, probability of survival, and man number of eggs laid by a female in its entire life span. He also made the logical point that, if the sex ratio were 1:1 each female would have to produce two young which survived to adulthood if the population were to remain in equilibrium.

Much of the early study of aquatic populations and the emphasis on the benthic and plankton communities was based on the idea of the "chain of biological relations" which was assumed to limit fish populations. The rationale for studying all the species which was evident in much early census and survey ecology derived, according to Baker, from the balance-of-nature concept as sated by Forbes. Nowhere can one see more clearly illustrated what may be called the sensibility of such an organic complex-impressed by the fact that whatever affects any species belonging to it, must speedily have its influence of some sort upon the whole assemblage. He will thus be made to see the impossibility of studying any form successfully out of relation to the other forms the necessity for taking a comprehensive survey of the whole as a condition to a satisfactory understanding of any part.

Forbes fell back on the familiar simile of clockwork, likening a missing species to leaving out one wheel of a watch. This mechanistic model is sometimes applied in ecology to illustrate the necessity of a holistic or hierarchical approach. Population surveys of many sorts of aquatic organisms proliferated. Somewhere in the development of fisheries biology the traditional holistic concept, so eloquently stated by Forbes, was lost. Fisheries biologists turned to an autecological emphasis on stock and production of single species, thereby divorcing themselves from major developments of ecology.

This is most concisely evidenced in a recent dispute concerning the relevance of ecological theory to fisheries biologists. Werner, in a number of book reviews, noted the largely independent development of fisheries biology and ecology and urged the necessity of basic ecological, particularly community, theory for fishery studies. Kerr rebutted, distinguishing between the synecological and evolutionary perspective of academic ecologists," as stated by Hutchinson and the "basic fisheries approach," or the "autecological perspective," set out by F. E.J. Fry. Fry's "perspective" was essentially similar Shelford's "tolerance" concept, extending the "zone of tolerance" into a surrounding "zone of resistance." As its defender, Kerr, and its critic Werner, agreed, Fry's concept stressed autecological response to physicochemical factors of the environment,. Hutchinson's theoretical niche was also related to Shelford's tolerance" concept, but incorporated biotic interactions or synecological approach.

The above exchange is useful in a number of ways. Kerr sharply criticized "academic ecology" and "academic theory" for its "parochial" perspective and failure to understand fisheries ecology. His distinction suggested the institutionalization of fisheries ecology in nonacademic, commonly governmental research organizations with different funding and purposes not shared by most academic ecologists. The American fisheries Society, organized in 1870, long antedated ecological societies, and later fisheries biologists were largely trained as specialists in fisheries departments or as systematists with little general ecological concern, according to Werner.

Werner argued that fisheries biologists should be trained as ecologists and that ecological theory, emphasizing interspecies interactions and community, is essential for effective management of fisheries populations. Fisheries ecologists had, however, developed theoretical population constructs of their own Rigler like Werner, took note of the minimal interaction between limnologists and fisheries biologists, ascribing it to their having quite different "paradigms." Rigler described the fisheries biologists' paradigm as stating that "the reproduction, survival and production of any fish population is virtually independent of changes in abundance of other biotic components." Rigler described the fisheries biologist's paradigm as stating that "the reproduction, survival and production of any fish population is virtually independent of changes in abundance of other biotic components. Limnologists, he said, continued the Forbesian view, and even went beyond it, in believing that "the appropriate system to study comprises the whole lake or stream and, more recently, even the entire drainage basin.'

Rigler urged convergence of these ideas in a return to empiricism, which, he argued, has been misunderstood and unfairly maligned. Actually, all the participants in this discussion deplored the independence of these different aspects of fisheries ecology and saw mutual advantage in improved communication. One, however, held out much hope for any proximate rapprochement, because of the long institutional and intellectual traditions which separate the two.

It is unfortunate for ecology that most of its concepts and theories developed largely in academic contexts, commonly divorced from the facilities, funding, and data available in government agencies concerned with natural resources. Although individuals employed by government fisheries agencies, forestry departments, biological surveys, ecological science was largely peripheral to the major concerns of most such agencies. A striking demonstration of the lack of recognition of professional ecology is that until 1977 there was no classification of ecologist in the U.S. civil Service although one had been proposed to Congress in 1920.

Early terrestrial plant ecologists, having the luxury of sessile and visible organisms, commonly estimated the abundance of species populations but, after 1898, also turned to counting. They were, at the turn of the century, favored in the taxonomy of plants was relatively complete and, then and since, by the fact that plants, unlike Elton's terrestrial animals, did not run away or hide at the approach of an ecologist. Early in the 20th century, counts of individual plants were supplemented by estimates or measures of area of cover, or basal area, which provided an index of size or quantity. Foresters had long since measured timber quantity or volume and growth, but such "mensuration" was usually restricted to commercial species and to usable timber rather than total production.

Clements advocated "clip quadrats" to measure production of herbage; and the pioneers of grassland ecology, such as Pound, Clements, and Weaver, and of its offspring range management. Such as A.W. Sampson, measured production of above ground herbage in relation to the environment and the effect of grazing. This is one of the few instances where there was effective interaction between agriculture and academic ecology. In most ways agricultural biologists, like fisheries biologists, did not develop close connections with ecology.

As John Harper complained many years later, population studies of crop and pasture plants and weeds done by agricultural scientists and range managers were largely ignored by later ecologists. The division was again largely in terms of autecology versus synecology. Agricultural biologists concentrated on growth and production of populations, whereas most ecologists sampled populations as members of communities. Terrestrial animal ecologists, by all accounts, including their own, were laggard in population and community studies, although a limited number of bird populations were censused in the 19th century.

As usual, S.A. Forbes was out in front with a survey of birds in a cross section of the center of the state of Illinois, showing their distribution in relation to various habitat types. W.L. McAtee did an intensive census of 4 square feet in which all plant and animal objects of classes known to be used as food by birds were counted." Forbes had stated that applied entomology was ecology long before ecology became a familiar term; but few applied entomologists seized on this observation to label themselves ecologists. Nevertheless, members of the long-established American Association of Economic entomologists were among the 307 charter members of the Ecological Society of America and entomologists were prominent among the leading pioneers of population ecology.

Among the earliest animal population studies of theoretical and empirical interest to ecology were those of the economic entomologist-parasitologists Ronald Ross, and L.O. Howard, then chief of the Bureau of Entomology of the United States Department of Agriculture, and his associate W.F. Fiske. Ross developed mathematical equations showing the relation between the incidence of malaria in humans and the population of anopheline mosquitoes.

From manipulation of these equations he stated as laws that the maintenance of the disease required a given number of mosquitoes and the beyond this maintainance level small increases in mosquito populations caused large increases in incidence of malaria. Howard and Fiske, though less formally mathematical than Ross in their treatment of parasites in controlling pest insect populations addressed a key aspect of population regulation. They distinguished between factors such as storms, which destroyed a fixed percentage of a species whether it was rare or common that is, independently of its abundance and "facultative" factors, such as predators or parasites, which, they said, were radically different. They wrote that

> a natural balance can only be maintained through the operation of facultative agencies which effect the destruction of a greater proportionate number of individuals as the insect in question increases in abundance.

The substance of their "proposition was that a parasite could control a pest organism only if the numerical increase of the parasite directly affected by the numerical increase of the host. Howard and fiske's proposition is an early, if not the first, statement of what came to be called the theory of density-dependent population regulation, which was to occupy the attention of animal ecologists for decades. Its longevity and contentiousness is suggested in Kerbs's lament that "the density-dependent regulation model is the ecologist's phlogiston theory, useful years ago but now an obstruction to progress." Oddly, although these significant developments in ecology were based on work with parasites, parasitology generally did not mix with ecology. Parasitologists were few among he charter members of the Ecological Society of America.

A continuing anomaly is that parasitologists generally remained aloof from ecology in its formative years and did not expressly couch their discipline in ecological terms although it is the most ecological of biological disciplines. Ecologists reciprocated by marked disinterest in parasites, preferring to concentrate on predation and competition as the significant two-species ecological processes. Ecologists per se were still a handful in the 1920 and 1930s. membership in the American Ecological society had by 1925, nearly 10 years after its founding, approached an asymptote, as the then newly rediscovered concept of the logistic curve said it should, of about 600 members, which lasted until 1950. The British Ecological Society around 1930 had about 450 members, largely botanists. It is a matter of some interest that animal ecologists in America were relatively numerous and active in the earliest years of ecology and in the formation of the Ecological society of America. The society's first president was a zoologist, and its first journal, Ecology, included plant and animal papers. The British Ecological Society was initiated primarily by botanists. Its nine presidents in the first two decades of its existence were botanists; and its first publication, the journal of Ecology, was largely botanical.

In 1977 the abstracts of a meeting of the newly formed Ecological Society of America were published in the Journal of Ecology with the remark that "it will be noticed that animal ecology occupies a prominent place in the programme. "O.W. Richards also commented on the neglect of animal ecology in Britain, as did Elton. Although plant ecologists were active in Britain, there wee no academic or government posts for plant ecologists as such, and growth of plant ecology was inhibited until the 1940s. animal ecology and limnology, were, however, gathering interest and support, as evidenced by the establishment of the Freshwater Biological Association in 1930 and the Bureau of animal Populations in 1932.

The Journal of Animal Ecology was also established in 1932, by Elton.: Prior to about 1925, relatively few ecological studies of terrestrial animal populations had been published. Lack in one of the first volumes dealing with regulation of numbers in natural populations of animals, referred to the despair of some ecologists of the population problem being capable of solution. "He reviewed the early long-term censuses of animal populations, largely of birds, relatively few of mammals and invertebrate animals.

The earliest, of a moth, started in 1880, and that of a bird, the great crested grebe, in 1860. Most early animal population records were made by amateur ornithologists, economic entomologists, or fisheries personnel for economic reasons, not primarily to resolve ecological questions, which were not very explicitly phrased in any event. The first decades of the 20th century marked the rise of statistics and its introduction into biology, including ecology. As early as 1970, Hjort urged the use of actuarial statistics in fisheries. Pearl noted the important contributions of statistical science to biology, namely (1) describing a group in terms of its own attributes, (2) estimation of probable error, and (3) measuring association or correlation between variations in a series of characters or events.

Ecologists of various persuasions had established the utility of describing a population in terms of its numbers of individuals, frequency of occurrence in samples, and productivity or biomass in both absolute and relative terms. Marine ecologists, such as hensen, had used percentage deviation of a sample from the mean, although not with an estimate of probability, as a measure of variation among samples. Ramensky and Gause used standard deviation in describing population variation.

Measures of association between species had been proposed by Forbes and were elaborated by Shelford and Michael. Michael reexamined Forbes's index and found wanting as a useful coefficient. His article voiced the dilemma, later to become endemic in ecology: Shall quantitative methods in biology be worked out by mathematicians, biologists, or both in cooperation? His solution was to require "proficiency" in mathematics of biologists.

As he put it "This means reform." The editor questioned whether the time was right to insist on proficiency in mathematics for biologists, although he allowed the need for adequate training in calculus and statistics. The full impact of mathematical and statistical ecology was in the wings in 1921, although many ecologists managed to evade it for some decades; and the perennial debate about the use of and need for mathematics begun by haeckel was to flourish in the 1920s and 1930s and be rekindled in the 1950s. Studies of what Park described as "quantitative census of natural populations," although made at various levels of accuracy and with limited check on the validity of the sample as a representation of the populations, indicate the interest of early ecologists in numbers of organisms. This interest was ubiquitous in all areas of ecology and gave rise to what Elton called the "statistics" of populations.

By this he meant the descriptive statistics of the population which he equated with numbers of individuals at one time. He contrasted it with "population dynamics" or change of populations in time. His analogy was with morphology, and he compared statistics to a still photography, and dynamics and development to moving pictures. This is a conveniently logical distinction, but not all population censuses were purely descriptive, although only a few concentrated on the elements of natality, survival, mortality, and dispersal or migration of populations, which were to become the essence of a new era of dynamic population ecology. Repeated and long-term studies of populations to follow changes were few, except in studies of commercially valuable species.

Early plant ecologists have commonly been criticized for their undue attention to description, qualitative or quantitative, of vegetation and their presumed lack of interest in dynamics of populations. Tensley wrote, "It is the fashion in some quarters to deride descriptive science as not science at all." Tensley commented that plant ecology should go beyond description to causation, a hope repeated nearly six decades later in an article entitled "After Description" by John harper, the leading contemporary student of plant population ecology. Harper exemplified the view that the causation urged by Tansley required

> concentration of effort on the lives and deaths of individual plants a reductionist approach as the most likely to reveal those forces at present operating to determine the distribution and abundance of plants.

It is somewhat ironical that the onus attached to description in Tansley's era was directed at verbal or natural history descriptions. By Harper's time the pejorative phrase "mere description" was largely intended for statistical descriptions of populations and communities, even when these were concerned with succession of change of the populations the community. Plant ecologists had anticipated Slobodkin's dictum that making fences around areas of the earth's surface and listing the kinds and numbers of organisms present constituted the beginning of population ecology. Sometimes the fences were literal for the purpose of excluding grazing animals.

More often the fence was abstract, enclosing an area of ostensibly homogeneous vegetation, and sampling smaller areas therein was taken to be representative of the whole. Although these ecologists usually failed to address questions of reproduction and survival of individuals, which be came the crux of population ecology, they did address significant problems about population. Not the least of these was how to "how much" The distinction of dominant and subordinate species and the propensity of a species to become dominant or relatively very numerous were major concerns. Thus studies of the distribution and abundance of plants served as a necessary preamble to or concomitant of the studies of lives and deaths of individual plants, as advocated by Harper. To explain how a population got that way, it was necessary to ascertain how it was. In spite of the preoccupation of most terrestrial plant ecologists with the community, for example, with vegetation, they necessarily were concerned with the plant species and the populations thereof. Like their aquatic ecologist counterparts, they counted numbers of individuals in areas and estimated quantity in a variety of ways, usually by projecting the plant area onto the ground. Terrestrial plant ecologists and early range managers at several experimental ranges in the western United States initiated extensive quantitative studies of plant populations.

Terrestrial plant ecologists, like made ecologists, commonly assumed homogeneity as an attribute of the community and its populations, often while paying little attention to whether this was true. Good all wrote that H.A. Gleason, in 1920, was the first to study quantitatively the spatial distribution of individual plants. Gleason was aware of the effects of quadrat size and number on estimates of population. He was among the first to apply probability to the examination of field populations and to devise an estimate of aggregation or clumping of a population using the mathematics of probability. The issue of population distribution. Whether homogeneous. Heterogeneous, random, or patterned, arose in ecology in the 1890s with Haeckel's criticism of Hensen's plankton theory. It continued and remains a stumbling block for population, community, and theoretical ecology.

Using the attributes of spacing, such as the mean area per individual or, more significantly, direct measurement of interorganism distance, was advocated in the early decades of the 19th century but rarely used until recently. If as Gleason thought, plant species were randomly distributed, frequency," the number of sample areas in which a species occurred, would have solved many problems of sampling. Because of the ubiquity of clumping in species distribution and the effect of size of sample area on frequency, it was inappropriate to use frequency as a refined estimator of populations. However, in conjunction with density counts, it did serve as a rough indicator for indicating homogeneity of distribution. The other use of frequency in quadrats was to determine co-occurrence of two species as a means of expressing association between them, a technique used earlier by S.A. Forbes.

Gleason recorded joint occurrences of pairs of species and compared then with the number expected on the basis of their individual frequencies, although he did not use tests of significance. Tests of interspecies association were slowly incorporated into quantitative plant ecology in the 1930s and 1940s and adopted by animal ecologists in the late 1940s.

Association between species is itself an indication of departure from randomness and suggests either biological interaction between the species or similar, or inverse, relation to the environment, depending on with the association is negative or positive. Quantitative measures of populations were also used to test correlation between vegetation and environment to examine the tolerance of plant species to environmental variables, particularly soil attributes. Many such studies suffered the usual problems of correlation without grounds for inferring cause and were often criticized as a descriptive-correlative approach to ecology. The literature of animal ecology in the 1930s included numerous studies denoting numbers of individuals of one or more species population with very little consideration of how they reached the observed number or if, or how, they might be changing.

When the Journal of animal Ecology was founded in 1932, nearly 70 percent of the articles published in it used quantitative methods; by 1970 this increased to 100 percent. Many studies which provided statistics of populations also suggested something of the factors involved in the dynamics of the population. Elton, like Shelford before him, joined plant ecologists in placing major emphasis on the community. andrewartha claimed that Elton predicted that "the laws governing the distribution and numbers of animals in nature would emerge from the study of communities."

According to Andrewartha, the converse is more nearly true: "We must first discover the laws governing the distribution and abundance of animals before we can advance very far in our understanding of the relationships between the populations that make up a community." In fact both Shelford and Elton turned their attention to populations, albeit in different ways. The balance of nature does not exist, and, perhaps, never has existed. The numbers of wild animals are constantly varying to a greater or less extent, and the variations are usually irregular in period and always irregular in amplitude.

In the process he attacked the ubiquitous clockwork simile, which had been used by Forbes and other ecologists, on the grounds that the animal "wheels" each have their own mainspring and retain the right to move to another clock. Elton emphasized the capacity of animals to migrate and described what he termed selection of the environment by the animal rather than the more usual selection of the animal by the environment. This had been emphasized may years before by Davenport. Studies of natural populations and of the problems of securing adequate samplings during the 1930s and 40s followed in the tradition of ecologists such as Elton and Leopold. Bodenheimer noted "the two greatest steps in ecological progress. The first puts numeric changes of populations in the center of all research; the second is the principle of compensation. Compensation is the tendency of a population to react to stress to maintain its numbers. Bodenheimer identified A.J. Nicholson and P.A. Errington as two figures connected with these steps. nicholson worked primarily with insects and stressed density-dependent control of populations. Errington worked with mammals and was an outstanding authority on predation but allowed that diverse mechanisms controlled populations. Bodenheimer illustrated the gap in animal population studies.

It has become commonplace to say that ecologists studied communities and animal ecologists studied populations. It is certainly true that terrestrial plant ecologists of the first half of the 20th century spent much effort in describing, classifying, and mapping communities by the credos of different schools. Their interest in community was supported by extended samples of the populations thereof and by various, often unsupported, inferences of the reasons why the populations changed. It is true that theoretical population ecology and some of the experimental work associated with it were done largely by animal ecologists or by nonecologists.

Terrestrial animal ecologists were preoccupied with numbers of organisms of individual populations and the changes therein, and they even went so far as to define ecology as the study of numbers of animals. It is a matter of some interest to inquire why this distinction should be so commonly asserted and to what extent it is true. Certainly some plant ecologists as early as 1919, notably A.S. Watt in his studies of woodland, were concerned with population phenomena. Watt's later studies of gapphase replacement and grazing by rabbits were also concerned with population phenomena. Both Clements and Tansley conducted experimental studies of competition, a favorite concern of animal population ecologists. Clements, Weaver, and Hanson produced the first major reference work on competition and offered an explicit definition of the term, while some animal ecologists, such as A.J. Nicholson still confounded its meaning.

Terrestrial animal ecologists may have turned to their traditional emphasis on single-species populations because of the difficulty of sampling animal populations as noted by Elton. Plant ecologists found it efficient to sample several species at once. Some plant ecologists found believed it was necessary to establish the community and environmental context of species so that autecological studies could be accomplished most efficiently.

The conspicuous variation of vegetation in form and composition evident to the casual observer, challenged plant ecologists to seek, or to provide, an orderly arrangement. The relative ease of collecting data on multiple species perpetuated the emphasis on plant communities, in part because there was so much variation.

During the 1940s and 1950s, new methods of sampling and data analysis focused attention on problems of sampling, patter, and association of species in communities, largely attended to by plant ecologists. The details of demography of individual populations or limited tow-species interactions, which intrigued animal ecologists, were largely ignored by plant ecologists, although Salisbury's work on plant reproduction pointed in that direction. Whatever the reasons, studies of plant population dynamics lagged. Harper wrote about a demography which has never "gained a momentum in plant ecology" and later said that plant ecologists "largely ignored population phenomena" E.O. Wilson even wrote of Harper that he was "the first plant ecologist to talk to talk to the animal ecologist in his own language of population ecology". Plant population biology was described in the 1970 as yet another "revolution and one of the paradigm changes which were epidemic in ecology. In fact, this revolution was described as "the most important event in the field of ecology" in the 1970s.

This particular revolution included the hope of merging evolutionary biology, physiological ecology, and population ecology in the 1980s. The essences of this hope was that detailed studies of population phenomena would provide generalizations which could be extended to encompass the phenomena of communities which had long engaged plant ecologists and which intrigued animal ecologists of the new theoretical ecology which had emerged in the 1950s. It is not now clear that this merger will bring together plant and animal population ecologist.

Schaffer and Leigh questioned the relevance to plants of population theory primarily based on animals. Harper also noted idiosyncracies of plants which suggest that observations and theories based on plant populations may not be readily miscible with those of mobile animals. Another concern of current plant population ecologists, which they share with turn-of-the-century ecologists, is the meaning and identification of adaptation. Harper suggested the word has lost its value. With all these difficulties.

Some plant population ecologists share with many animal population ecologists the belief that "ultimate ecological explanation" must derive from studies of populations on the "lives and deaths of individual plants" a " reductionist approach". The 1920s can be seen as a watershed in animal ecology because the decade produced a significant codification of what ecology was all about in Elton's Animal Ecology. Essentially simultaneously, mathe-matical population theory was formulated and served, according to later mathematical ecologists, to add insight to most of the ideas Elton had outlined as the essence of ecology.

The dichotomy between those who saw the community, or later the ecosystem, as the ecological theater in which the play of populations took place and as the key to understanding the functions of the parts, or species, and those who claimed that it was necessary to comprehend the working of the species populations, as a preamble to understanding of community, really began in the 1920s with the formalization of theoretical mathematical population ecology.

Theoretical mathematical ecology was accompanied by increased study of experimental populations in the laboratory, thus completing the triad of "modern population studies-natural populations," "experimental populations," and "theoretical populations" described by Park. This set the stage of the ecological theater for extended discussion about the relative merits of each part of the triad and the relevance of each to each which continues to he present.

The tradition of census or enumeration of aggregates of natural populations of organisms, which had been one of the forerunners of self-conscious ecology, persisted particularly in benthic and plankton studies and to a lesser extent in terrestrial animal population studies. An interesting view of the continuance of the link between community and population studies is found in Elton and Miller in a review of survey of animal communities:

> As of the main class of biotic relations... are between different species, it follows that some reconnaissance of species associations in nature is needed in order to under stand the population ecology of any one of them.

The 1920s marked the beginning of theoretical mathematical population ecology its "Golden Age," according to Scudo and Ziegler Nowhere in ecology is the contrast between the inductive empirical and deductive/theoretical approaches more marked than in the traditions of census of natural populations and of models of abstract populations or experiments with markedly constrained populations designed to test mathematical models and theories. fortunately, several recent considerations of the history of ecology are concerned with theoretical population ecology and its earliest and omnipresent model, the logistic curve.

The potential of populations to grow geometrically had been recognized by Thomas malthus at the end of the 18th century. Interest in the mathematics of population growth was evident early in the 19th century, notably in the form of the logistic curve formulated by Verhulst in 1838, which went largely unnoticed. Early in the 20th century, mathematical models of growth of individuals intrigued a number of biologists, among them the zoologist-geneticist Raymond Pearl.

Pearl's interest in human populations was stimulated by working for Herbert Hoover in the Malthusian context of Europe at the end of World War I. Two concerns manifest at that time were population pressure on the limited means of subsistence and population statistics. In 1920 pearl and L.J. Reed rediscovered Verhulst's unheralded logistic equation with quite different result due to Pearl's effective public relations campaign supporting it as a "law" of population growth.

Pearl arrived at conception of the population as a whole. He made the point explicitly in 1927 that "populations of whatever organisms are, in their very nature, aggregate wholes, and behave in growth and other ways as such". The holistic concept which traditionally permeated community ecology was thus explicitly introduced into theoretical population ecology. Pearl and Reed recognized two key assumptions of the logistic equation: (1) the rate of growth of a population is proportional to the number of individuals at any time, and (2) there are still unutilized potentialities of support existing in any given area. The logistic curve describes the growth of a population over time.

Graphically, it takes the form of an S-shaped or sigmoid curve. In 1921 Pearl invited A.J. Lotka, who had been trained in physics, to work in his laboratory. Lotka had for some years been concerned with evolution, population growth, and mathematical formulation of these, but he did not become interested in the logistic equation until around 1923. Lotka incorporated a less dogmatic view than Pearl's of the logistic equation into his book elements of Physical Biology one of the landmarks of population theory. In it he drew attention to Ronald ross's malaria equations, noting the similarity of ross's curve to the logistic. He also reviewed W.R. Thompson's efforts to analyze mathematically the effect of a parasite on its host. finding Thompson's formula inadequate, he developed paired differential equations, showing the effect of a parasite on its host. These equations produced cyclic fluctuations of both populations.

Lotka proposed studies in "general demology" and reviewed some field studies, particularly of aquatic organisms, which, he said, were facilitated by the three-dimensional aquatic system and their connection with practical fisheries problems, which furnished "financial support on an extended scale," an unusual characteristic of ecological studies until every recently. His own work concentrated on the abstract one or two species cases. Independently of Pearl and Lotka, Vito Volterra, a mathematician, arrived at equations for two-species interactions. Volterra had early interests in mathematical biology, but his ecological interest came by marriage of his daughter to a zoologist who was interested in fish populations.

Volterra's attempts to explain mathematically the balance of nature started with the logistic equation, arrived at by methods different from those of Lotka, and equations of two-species interactions. Hutchinsion and Deevey wrote of these:

> Perhaps the most important theoretical development in general ecology has been the application of the logistic by Volterra, Gause and Lotka to 2 species cases.

Somewhat overlooked in the subsequent widespread application of the logistic equation and its extension to the two-species interactions competition, predation, and parasitism was Volterra's claim that he had treated as "new" the application of the 'fundamental laws' of population ecology to the "cohabitation of *n* species," that is, to the community. Volterra's *n*-species models were largely ignored, only to be rediscovered in the 1950s. Also lost in the general isolation of population from community ecology was the fact that Lotka and Volterra were after bigger game than populations.

As Leigh emphasized both Lotka and Volterra focused on the community, characterizing it as a state vector with as many components as there were species in the community. Lotka was far more ambitious than Volterra. Among at a kinetics of life, consciousness, and human social organisation, a reach not seen again until

H.T. Odum's extensions of aspects of Lotka's work. Whatever the criticism of the logistic equation, mathematical population ecology, initiated by it. Became and continues as one of the most active areas of ecology. According to Hutchinson and Deevey Population biology is the only part of biology other than genetics to have developed an autonomous quantitative theory independent of the physical sciences. It may be questioned how autonomous or independent of the physical sciences mathematical population theory is. Some of its major proponents. form Lotka and Volterra to Robert May, have been mathematicians or physicists, and much of its rationale is explicitly derived from physical science. This is a virtue in the eyes of those seeking to turn ecology into a "hard" science and a vice to those who deplore "physics envy."

Kingsland traced the explicitly physical-chemical analogy to individual growth, and he compared density effects to kinetics of gases. Lotka treated the organic and inorganic as a single system and analyzed the population as a chemical system with exchanges of energy and matter as the criteria. He drew largely on the second law of thermodynamics and treated organisms as engaged in competition for energy. His grand thesis was that evolution would increase the flow of energy and matter through the system.

Kingsland noted that the audience which Lotka hoped to reach consisted of physicists and it was C.C. Adams who pointed out to him the connection of his work with ecology. Volterra used the same physical analogies as Lotka, if less literally, but was really concerned with a mathematical problem. The entomologist A.J. Nicholson developed a verbal model of two-species population interaction and induced a physicist colleague, V.A. Bailey, to provide a mathematical formulation.

Nicholoson assumed a machine-like balance in nature, and Bailey's proof treated the organisms as gas molecules. Nicholson provided "a simple analogy" of a population in balance with its environment like a balloon as the environment fluctuated, rising and falling in accord with the gas laws. The original mathematical formulations generally assumed all individuals were identical and the environment was homogeneous and constant. Criticism of mathematical population theory began early and continues to the present. Charles Elton, in a review of Lotka's book, wrote a common complaint of many later ecologists:

Like most mathematicians he takes the hopeful biologist to the edge of a pond, points out that a good swim will help his work, and then pushes him in and leaves him to drown. Nice expressed the concern, common to many ecologists working with natural, or even experimental, populations that the diversity and complexity of natural populations could not be effectively expressed in simple mathematical formulations. she allowed that the fault had to be shared by naturalists in not giving theorists the data on which to work. She joined other ecologists in emphasizing the need for "a great body of facts intelligently and conscientiously collected before we can safely launch into elaborate theories.

Ironically, Thomas Park in a discussion following his review of theoretical and experimental population studies, commented that even Nice's unusually long study was not long enough to indicate stability of populations, and thus posed a familiar problem of natural populations as applied to theory. Theoretical mathematical ecology of the golden Age made headway in spite of the reservations expressed by many ecologists. In 1949 Park joined his distinguished associates in commenting that "theoretical population ecology has not advanced to a great degree in terms of its impact on ecological thinking. "Nevertheless they saw "bright promise" in experiments "based on theory". F.E. Smith commented on theoretical population ecology's "profound effects on applied fields," but he described the several mathematical formulations widespread in the field a s hypotheses. In Smith's view. "With the possible exception of the concept of exponential growth, it is a misuse of terms to refer to any of the interpretations as 'theories." at the threshold of the resurgence of theoretical mathematical ecology in the 1950s Smith wrote,. "The degree of acceptance of such concepts as, for examples the Verhulst Pearl logistic and the Lotka-Volterra equations, is astonishing."

Clearly, in current ecological jargon, these mathematical formulations and the assumptions about science on which they rest, are resistant," "resilient," and "persistent." Efforts by ecologists to make them conform more closely with biological reality began with G.E. Hutchinsion's addition of a time-lag term to the logistic equation. Watt reviewed population theory and noted that the conclusions of Volterra, Lotka, and Nicholson and Bailey deductions from their assumptions but that "the assumptions are not drawn from biological reality."

Current volumes of theoretical ecology still begin with these workhorse equations, as amended and expanded, and review progress in their use often with little reference to early and continuing doubts concerning their utility in providing explanations or predictions. May however, reviewed the conundrum of the classic deterministic logistic equation producing random dynamics rather than a smooth approach to the asymptote. In may's words, "to see mathematical ecology informing theoretical physics is a pleasing inversion of the usual order of things."

The real issue is, of course, how well it informs ecology Slobodkin provided a list of do's and don'ts for mathematicians in population biology. One which is germane is the tendency he noted of mathematical theorists to dodge criticism. Slobodkin suggested that mathematicians learn biology and then create new and appropriate mathematics rather than fitting poorly understood biology into mathematical molds which were created for hydrodynamics, thermodynamics, economics, or other concerns.

Theoretical ecology, competition, and equilibrium

Modern population ecologists conventionally interpret the history of theoretical population ecology as if theory led to the idea of competition and its use as a basis for community organization and niche theory. G.E. Hutchinsion said that Volterra's mathematics stimulated Gause and that from Gause's law. Diamond asked a related question in an article entitled "Niche Shifts and the Rediscovery of Interspecific Competition: why Did Field biologists So Long Over look the Widespread Evidence for Interspecific Competition That Had Already Impressed Darwin?" The correct answer was that they had not overlooked it, and Diamond's s several reasons for his assertion that they had were effectively refuted by Jackson made the point that ecologists, particularly plant ecologists, "did not need a mathematical formulation of the niche... to have the same idea", and they had, he asserted, "a well developed theoretical framework, namely succession before formal population theory existed.

Jackson reviewed the history of the concept of competitive exclusion and niche theory in American and British ecological journals and concluded that much work on interspecific competition antedated or developed independently of the mathematical theories of Lotka,

Volterra, Gause, and their successors. He observed that mathematical population theorists ignored the accomplishments of earlier ecologists: Since many recent proponents of theoretical ecology see "modern" ecology as primarily stimulated and led by theoretical ecology both in the 1920s-1930s and the 1950s--1960s, Jackson's evaluation of the way ecology progressed in respect to competition in a useful and timely caveat which warrants examination.

F.E. Clements and his associates J.E. Weaver and H.C. Hanson, in the earliest treatise on competition reviewed early ideas about competition in the work of Malthus, DeCandolle, Darwin, and Nageli. Warming reported the restriction of plant species from certain habitats by competition, forcing the species to grow in another habitat, which gave it the appearance of preferring that soil. He anticipated the interest in competition of later generations of ecologists: "There is scarcely any biological task more attractive than that of determining the nature of the weapons by which plants oust each other from habitats. "Experimental studies of plant competition began, according to Clements, with his own work in 1903, which included both field and greenhouse experiments. He noted the consequences as numerical reduction or disappearance. He also cited Tansley's inference about the difference in habitat of species of plants when grown together and alone. This is one of the earliest experimental examples of what came to be called niche separation.

Plant ecologists studied the relation of competition to succession, and a great deal was made of the "complementary" status of species in occupying different soil depths, seasonal periods, or aerial strata as a consequence of evasion of competition. some of these studies were guilty of inferring competition from the fact of segregation a guilt which they shared with a very large number of subsequent studies of niche segregation of animals which also assumed competition was the cause of spatial, habitat, and morphological separation of species. The significance of competition was clearly stated by early ecologists: "competition exerts a controlling influence on succession"

Although terrestrial plant ecologists had the advantage that plants stood still, they also had the difficulty that the different parts, root and shoot, they also had the difficulty that the different parts, root and shoot, existed and interacted in different media, creating the problem of differentiating root from shoot competition. Among the other observations of early plant experimentation was that competition between species might be masked by grazing of the dominant species. Much of the work of plant ecologists on competition was concerned with invasion or replacement of species through generations in the process of succession.

Given this interest, it is striking that the experimental demography of plant populations was not pursued in the 1920s and 1930s, when animal demography came to the fore. Although plant ecologists and foresters were concerned with age structure, commonly inferring it from size classes, the first actuarial analysis of a plant population in the sense of a life table was a study of *Plantago lanceolata* in 1959. Experimental studies of the dynamics of single-species populations and two-species populations of animals did develop in substantial part in association with mathematical theory. Pearl began studies of Drosophila to study the effect of density on fecundity and population increase. Pearl and several associates studied mortality and survival and urged careful study of cohorts and the use of life tables, but as Deevey put it, ecologists were busy elsewhere." Nevertheless, a tradition of experiential population ecology was established.

An important sequence of this tradition began with studies by chapman using flour beetles and continued in the notable experimental studies of Thomas Park The earlier hope faded that experimental studies or studies of growth of natural populations would validate the logistic equation as a "law," but the "heuristic" merit of mathematical theory was established as it stimulated much discussion of populations and life history phenomena. Gause called for experimental studies of the "elementary processes" of the struggle for life, concentrating on competition to such a degree that the theoretical effect of competition, the exclusion or death of one of the competitors, came to be known as Gause's principle".

Gause noted the early experimental studies of Tansley and those reviewed by clements, Weaver, and Hanson but asserted, inaccurately, that they were concerned with ontogenetic development of individuals rather than with studies of replacement of one species by another over a number of generations. He also noted that there were few studies of animal competition to provide exact data on competitive exclusion or replacement of one species by another. Gause clearly defined the elementary process of "competition for a common place between a small number of species in a limited microcosm."

In his experimental work, Gause elaborated on such elementary processes as population growth, competition, and predation; but be joined Lotka and Volterra in emphasizing that his eventual goal was the study of the entire community. In spite of his eventual goal was the study of the entire community. In spite of his emphasis on populations, he was a proponent of the holistic ecology which dominated ecology. Gause even adopted some of the criteria of organicism as identified by Haraway Gause wrote:

It would be incorrect to fall into an extreme and to consider the complicated phenomena of the struggle for life in nature as simply a sum of such elementary processes... The elementary processes of the struggle for life take place amid a totality of most diverse living beings. This totality presents a whole, and the separate elementary processes taking place in it are still insufficient to explain all its properties. It is also probable that changes of the totality as a whole put an impress on those processes of the struggle for existence which are going on within it. Gause invoked the whole-and-parts relationship which has pervaded ecology from F.E. Clements to E.P. Odum. He allowed that there was a difference between the whole and the sum of its parts, which is often and variously stated but never effectively quantified. However, he likened community ecology in his day to 19th century biophysics, which he said made progress by fruitful studies of the parts or elementary processes... and thereupon only did the question arise of studying the organism as a system constituting a whole."

Gause's analogy was questionable, for natural historians, protoecologists, and "self-conscious" ecologists had been trying to study the community, with the assurance that it was a whole and had to be studied as such for some decades; and formal theoretical study of the parts and their elementary processes was only beginning in the 1920s and 1930s. Gause's hope for the utility of the study of simple experimental populations to extrapolate their lessons to complex natural communities was the basis of extensive studies of diverse taxa in the following decades, and many assert that such studies are the best or even the only way to real understanding of communities.

Leigh perhaps prematurely, described this program as a failure,...for the curious reason that simple experimental communities are too unstable and erratic to exhibit lawful behaviour." May, the leading modern expositor of theoretical population ecology, wrote:

> For communities with many species, any description of the population dynamics of individual species in terms of the interactions within and between them usually is very difficult.

The substance of the difficulty is that the mathematical models of single populations and two-species mixtures are not readily extended to more complex mixtures and to situations in which the biologically unrealistic assumption of the logistic equation and its derivatives are not valid. May's alternative was to abandon the analysis of populations and to focus on collective or aggregate attributes of communities such as energy flow, trophic organization species area relations, and diversity. Like some later animal ecologists, he confidently regarded the structure of communities as consisting of definite elements in a fixed numerical relation with each other." He quoted Elton's remark that the total number of species in an association is fixed and determined by some important principle. Gause's assurance that the organized community was an entity led to his statement of the one-species-one-niche concept decades before its reappearance in the heyday of niche theory in the 1970s.

Gause also addressed the question of the association as a distinct combination of species sharply separated from other associations and he noted that most botanists treated associations as relatively distinct he was one, however, of the very few ecologists who cited Gleason's individualistic concept, which was an almost lone voice of dissent from the conventional organismic community-unity concept in the 1920s. Interstingly enough, Gause did not cite his own countryman, L.G. Ramensky, who was enunciating the same criticism as Gleason of the then conventional wisdom about communities.

Gause's review of the principles of biocenology is useful in that it addresses most of the key questions which were and are intriguing ecologists such as pattern, the relation of density and frequency, the relation of number of species and numbers of individuals, and the stability or equilibrium of the community as related to these elements of structure. Gause recognized two types of community.

Unstable and stable or mature. According to Gause, in the early stages of development of the biocenosis all of the principal species grew rapidly, but later a few species exercised dominance due to "biocoe-notic relations." suppressing most other species. Gause, more confidently than later ecologists, wrote of lakes, as biocenoses. "It can be said with perfect confidence that the biocoenosis is subject to regular development and attains a certain stability," although he allowed that in temperate fresh waters a final stabilization of the system is never attained.

Gause summarized his review by noting two fundamental problems raised by field studies: (1) the regulation of stable combinations of species and (2) the separation of the biocenosis into individualized natural units, or types, in spite of an uninterrupted change in the external conditions. Gause noted the gap between mathematicians and ecologists and wrote:

> Mathematical investigations independent of experiment are of but small importance due to the complexity of biological systems. narrowing the possibilities of theoretical work here as compared with what can be admitted in physics and chemistry.

Nevertheless, his assurance to ecologists was that elementary processes can be successfully analyzed with the experimental method and "accounted for theoretical with the aid of a differential equation."

Thomas Park reviewed the development of analytical population studies and their relation to general ecology. Park, then well established on his distinguished career in experimental studies of animal populations, endorsed the logistic equation with reservations as a "valuable demographic tool." but noted conflicting experimental results. He remarked on the interaction of mathematical equations and experimental populations but said of population mathematicians that contrary to their avowed intention of elucidating the community:

> To date they have not added much material of tangible value to community analysis. This is attributable to the fact that their concern has been of necessity with over-simplified situations. Kingsland commented that Park, although he had spent the logistic equation was either a law or predictor of growth. However, she noted his section on population growth in Allee *et al.* stressing the advantage of the logistic as a demographic tool was important in the definition of population ecology as a discipline.

The idea of competitive exclusion, long familiar in ecology, formalized in mathematical population ecology examined in minute detail in experiment, and commonly inferred from distributions of natural populations, became and continues as a cornerstone of population and community ecology. Clements quoted Darwin's law of competition" as fundamental:

> As species of the same genus usually have, though by no means invariably, much similarity in habits and constitution, and always in structure, the struggle will generally be more severe between them if they come into competition with each other, than between the species of distinct genera.

A widespread assumption derived from this type of reasoning, was that closely related species would not likely be found in the same area. This was a concept with a long history J.B. Steere, studying bird distribution in the Philippines, observed "Every genus is represented by only a single species in any one place." Plant biogeographers and ecologists, notably Paul Jaccard, studied the ratio of genera to species which approached one and concluded that like elements were eliminated. David Starr Jordan the most prominent icthyologist of the era, asserted what some came to call "Jordan's law":

Given any species in any region the nearest related species is not likely to be found in the same region nor in a remote region, but in a neighbouring district separated by a barrier of some sort. In 1944 the British Ecological Society addressed this question in a symposium" The Ecology of Closely Allied Species." centering on Gause's contention that two species with similar ecology cannot live together in the same place". The revealed itself which can only produce a feeling of deja vu in an ecologist active in the 1970.

At the symposium, Elton, although he had expressed doubt about the significance of competition in Animal Ecology, jointed David Lack and G.C. Varley in supporting Gause's principle. Lack in his autobiography, wrote that this was "almost the only time in our lives..... Charles Elton and 1 were on the same side. "Lack based his defense on observations of niche differentiation and ecological release or expansion of habitat, when the competing species was absent or removed.

Varley emphasized the evolutionary impact of mortality associated with competition and argued that a consequence of competition would be adaptive radiation producing for example, differences in feeding habits and, presumably, structures Elton argued from community composition that a high proportion of genera are each represented by a single species, Unfortunately of the papers opposing Gause's principle were subsequently published, not was the lively discussion in which "arguments pro and contra were fairly evenly balanced" ever published Dr. Blackman argued that exclusion did not always occur, Captain Diver and Mrs. Spooner asserted that the mathematical and experimental approaches were dangerously over simplified."

Diver raised the difficult problem of defining "similar ecology and asserted that direct evidence of competition between species was lacking be cause other factors kept populations below the level at which competition would occur, Caswell noted that all of the arguments at the symposium have been elaborated continuously during the next 30 years. He also commented that Diver raised the issue. Which has again been raised by critics of more recent expositions of competition as a basis of community theory that other factors precluded the equilibrium conditions required by competitive exclusion theories.

Hardin commented that Gause's principle was born at that meeting rather than conceived because it had been gestating in the ecological literature for an extended period whose depended on the moment of. Jackson raced its conception to an earlier era of plant ecology. According to Jackson, Hutchinson and Lack, in their respective considerations of competition and niche in the 1940s misinterpreted the literature by ignoring terrestial plant ecology. Competition in ecology, Jackson said was biased by its equation with the formal mathematical theories of volterra, Lotka, and Gause, and all questions about competition were equilibrium case."

Oddly, although Hutchinson had described plant ecology as "arthritic," he did link the logistic curve with the growth of a biocenosis of the development of a sere" in a word, succession. He emphasized hat the components of a sere do not have genetic continuity but endorsed the then conventional view that pioneer " play an essential part in preparing the ground for qualitatively different successsors and modify the substrate to support a quantitatively greater flora." Hutchinson also commented that takes developed rapidly from oligotrophy to eutrophy and a long period of equilibrium in a trophic states Moreover, he said the growth curve of the lake resembled that of a single individual of a homogeneous population and was characteristic of a biocenosis, in general coinciding with Gause's view.

Although Gause made major contributions to ecology in the 1930 the principle attributed to him, according to Hardin, remains a potent part of modern ecology. K.E.F. Watt made it into two principles. The principles became the subject of continuing dispute, as Hardin noted, because of an inherent problem of circularity. When Ayala reported an experiment demonstrating that competing species could exist in a stable equilibrium. Gause came out of ecological retirement to deny the inference, on the ground that the two presumbly competing species actually were occupying different riches. Ayala responded, defending his invalidation of the principle of competitive exclusion but noting the perennial problem off inadequate definitions of competition and niche. In effect, this exchange indicated the circularity problem noted by hardin. neverthless, Hardin had asserted confidently concerning the history of ecology and the study of population genetices:

> We stand at the threshold of a renaissance of understanding a renaissance made possible by the explicit acceptance of the competitive exclusion principle.

Much of theoretical ecology from the 1950s to the 1970s rested on the competitive exclusion principle but whether it constituted a renaissance of understanding is still in dispute The ubiquitous attribute of balance stability or equilibrium arose and persisted in population ecology, as if had in community ecology. Biologists and ecologists involved in general survey or census of plant and animal populations had shared the belief that were in, or would tend toward, a stable or equilibrium number in the absence of human disturbance.

Mobius described a "biocoenotic equilibrium of a mean number of individuals of a biocenosis. S.A. Forbes described a benign result of the struggle for survival as an equilibrium of the interacting populations. In Forbes's dynamic view, populations oscillated, but the oscillations were kept within bounds and tended toward equilibrium. Adams recognized a "biotic base, optimum, or balance toward which relations under given conditions tend, at a time which an equilibrium will be established. "Shelford described the community as a balance in the numbers of contending organisms.

Chapman characterised the concept of equilibrium as one of the most fertile in the field of ecology. He extolled the theory of biotic equilibrium being developed by Volterra but noted tat a in addition to discordance of the physical environment as cause of population fluctuations the biotic system may generate its own fluctuations. Elton considered mechanisms which bring about the "optimum density of numbers for each species but noted that no population stayed the same for any great length of time". H.S. Smith described population equilibrium in the context of the new population theory in-term of the ate of change being approximately zero. Allee *et al.* accommodated Elton's assessment of constantly varying populations by suggesting a comparative balance in nature based upon long-term populations relations. How their long term related to

Eltons great length of time was not clear. Slobodkin commented that a "sufficient length of time" was a necessary condition for population equilibrium because the population had to be in an " equilibrium age structure. Diverse views no population equilibrium and the relevant terminology persisted. Bodenheimer attempted to clarify the concept of balance or equilibrium in his turn nothing that the meaning differed from that of a physical equilibrium.

Ehrlich and Birch described the idea of a balance in respect to population size as "demonstrably false." The 1960s and 1970s produced extended discourse about the meaning of stability and heated discussion of equilibrium and none-equilibrium theory. Council and Sousa returned to square one with an article "*On the Evidence needed to Judge Ecological stability or Persistence*" and the advice that one should determine if a population was stable before applying a theory based on premise of equilibrium."

The tradition of balance or equilibrium so strongly imbedded in natural history and early ecology led to varying assumptions about the factors responsible for maintaining equilibrium or for controlling populations in the absence of equilibrium. Clements assumed that populations were replaced until a relatively stable "climax" was reached in which the populations were self-perpetuating and stable.

Among the earliest explanations was that offered by Howard and fiske in a study of the introduction of practises to control pest insects. They recognised natural balance of a population as maintained by the operation of facultative agencies which effect the destruction of a greater proportionate number of individuals as the insect in question increases in abundance " This classic statement of density-dependent biological controls they said, led to considering parasitism as 'the most subtle in its action in controlling a population. The theory of populations being in or tending toward equilibrium was questioned by Uvarov as "directly country to the facts." He jointed Elton in questioning the balance nature concept and denied that a relatively stable normal umber or uniform fluctuations around it, characterized populations. He also question the primary of biotic factors in maintaining balance. The key to population control, he believed, was in the influence of climatic factors. It is fitting that this dichotomy was expressed studies of insects, because entomologists have bee prominent in the development and polarization of concepts of equilibrium and factors controlling populations. The entomologist Chapman, oddly enough even offered reasons for considering insect ecology apart from animal ecology.

The experimental studies of Chapman, pearl and Gause were take by the entomologist A.J Nicholoson as evidence for the existence of balance or equilibrium of animal populations in nature. Nicholoson jointed forces with Bailey, a Physicist to investigate the problem of competition and the relation between animals and their environment and considered that observation and experiment supported the concept of balance.

Nicholson asserted that it was usual at present to deny..... the existence of the balance of nature. "in fact, it wads widely assumed that balance was characteristic and Nicholson cited only uvarov and bodenheimer to the contrary. Nicholson believed that animal populations are normally in eqilibrium and fluctuated only within restricted limits. He used a gas-law analogy of a balloon to illustrate how a population would change if its environment changed but would rapidly come into equilibrium with the new environment, presumably as it stabilised.

Moreover, he argued that the balance of populations is maintained by a actor which is governed by the density of the population controlled and the asserted that "competition seems to be the only factor that can be governed in this way." Although competition had long been assumed to be a major factor in limiting populations and in organizing community structure. Nicholson's assertion of the so-called density dependent population regulation exacerbated a dispute which persisted for decades. Nicholson somewhat dogmatic assertion of the primary of competition and his extensive experiments attracted much attention.

Although his definition of competition was unclear, his position was that "balance is produced and maintained by competition. In fact his idea was somewhat circular. He said that competition is capable of producing balance' and the that any factor that produced balance is almost necessarily some from of competition." Among these factors he included limitations of things required for existence and ease of being found by natural enemies.

Nicholson's emphasis on mathematics was not acceptable, to some population ecologists. W.R. Thompson, who had been much taken by mathematical ecology in the 1920s, became much opposed to Nicholson's ideas in the 1930s. Thompson pressed the familiar argument that population biology was comprised largely of unique and complex events so that a general theoretical mathematical model was not possible.

Two landmarks of population biology and of the dispute concerning density dependent control of populations appeared in 1954. Lack attempted a general assessment in a book The Natural Regulation of Animal Numbers. Lack supported Nicholson's idea of density dependence, although he attributed it to a complex of interactions of food shortage, predation, and disease. Lack was explicit in his recognition of competition. He inferred it from differntiation of species in food habits, habitat use, and morphological displacement. Andrewartha and Birch flatly disagreed with the concept of density dependent control. They held that species fluctuated irregularly and were controlled, but not in any sense of balance or regulation, by factors, largely weather, which acted independently of the density of the population. As late as 1958, in the face of the substantial dispute, Nicholson asserted:

> Logical argument based on certain irrefutable facts shows not merely that populations may regulate themselves by density induced resistance to multiplication but that this mechanism is essential.

The problem in the 1930s, in the 1950s, and revived again in the 1980s was not simply that individual populations were to be understood as in or near a stable equilibrium, but whether a community "was best understood as assemblages of populations at or near a stable competitive equilibrium". Caswell noted that Nicholson, who was much influenced by the theoretical mathematical work of Volterra and Lotka, and David Lack, who was not, were among the supporters of density dependence. They commonly argued, as had Darwin, that populations had the capacity for exponential growth or for extinction.

To maintain a population somewhere between these levels the controlling factor must limit growth at high densities and enhance it at low densities. The perennial question of density-dependent and density-independent control of populations persisted, often confounded by misinterpretations of what these terms mean. Milne, for example, commented that his definition of population control was factual and "not confounded with theory as to how it comes about, citing Nicholson's density dependent control by competition as culpable in this respect. Nicholson's theory was widely accepted but was also attacked particularly by the entomologists Andrewartha and Birch.

Andrewartha and Birch, in a history off insect ecology, reviewed this dispute and argued that the probability of a population reaching a ceiling which provokes a density-induced resistance to multiplication" was low, and so was unimportant to population control. This argument turns on the assumption of equilibrium, and Andewartha and Birch commented that the controversy awaits a philosophical resolution. " As with many other philosophical discussions, the resolution was a long time coming. The discussion persists in current disputes concerning the equilibrium status of populations and communities.

Population ecology was considered at a symposium in 1957 bringing together human demographers and animal ecologists, with little consensus on theory evident. A notable product of this symposium was Hutchinson's "Concluding Remarks", which is certainly on of the least explicit titles in the history of population ecology, masking as it does one of the most highly touted and disputed productions of that discipline-a formalization of niche theory. Hutchinson's description of the symposium as " a heterogeneous Unstable population" is however, particularly apt. He contrasted the comment of one participant about the uselessness of the logistic equation with the almost universal practice of animal demographers to start thinking by making some suitable, if almost unconscious, modification of this much abused function.

The extension of this tradition to the present is noted by Sale, who commented:

> A fair statement of the current consensus is that the tropical community is a mature equilibrium community of numerous species whose coexistence is satisfactorily explained by the theory based on the Lotka-Volterra competition and predation equations.

Hutchinson also asserted, concerning populations, that a difference in interest may underline some of the arguments that have enlivened, or at times disgraced, discussions of this subject. He noted that laboratory workers in their experiments tended to keep all but a few factors constant ant to vary these few systematically, whereas field workers tended to emphasize the changing environment. Not specified was that theoreticians ignored any variation in the environment, assuming that it was constant and homogeneous. Hutchinson's salient comment was;

> The initial differences of point of view are not the only difficulty... It is also very likely that the differences in initial point of view are often responsible for the differences in the interpretation of the data.

The burden of Hutchinson's "Concluding Remarks" is a suitable place at which to terminate this chapter. He provided a formalization of the niche based on the competitive exclusion principle. Niche was the expression of the response of a population in an n-dimensional hyperspace in which "all X variables, both physical and biological", were to be considered. The introduction of the concept of the multidimensional niche of a species parallels the concurrent introduction of the use of multivariate methods in plant community ecology by Good all and bray and Curtis.

Ecologists, of course, had long since though of the entities which they studied as excessively complicated, perhaps too complicated for the human mind to grasp. Multidimensional analysis seemed admirably suited to treat the acknowledgedly multiple factors affecting a species or the multiple species constituting a community. The effort to bring them together in a unified theory of population and community by means of multivariate analysis essentially begins in the mid-1950s. Populations are fundamental units in ecology, as important to the ecologist as are tissues and organs to the anatomist and physiologist. They may be thought of as major components of communities and ecosystems, just a tissues and organs are major components of individual organisms while tissues and organs are composed of individuals in functional groups.

A population is a group of living individuals set in a frame that is limited and defined in respect to both time and space. Also population is an interacting, or potentially interacting group of individuals usually of the same species. It may be the population of ants in an ant hill, the population of salmon in a river, the population of a deer in a state and so forth. Biologically, a population should have some natural boundaries as is the ants in an ant hill, but the term is often used with artificial boundaries in mind, as in referring to the population of deer in a state.

Population attributes

In the initial study of populaion it is helpful to recognize certain attribute of population which distinguish them from individual organisms like organisms have a structural organization, a functional unity, and a pattern of growth and development unlike individuals they have group attributes and statistical properties that no single individual possess. The more important of these population characteristics or group attributes are as follows,

Density: Population size in relation to unit of space.

Birth rate: or more broadly natality the rate at which new individuals are added to the population by reproduction.

Death rate: or mortality rate: The rate at which individuals are lost by death.

Dispersal: The rate at which individuals immigrate into the population and emigrate out of the population.

Population growth rate: or growth form. The net result of natality mortality & dispersal.

Dispersion: The way in which individuals are distributed in space, generally in one or more the following patterns:

i) Random distribution, in which probability of an individual occurring in any one part is the same as the probability of it occuring at any other part.

ii) Union distribution, in which components occur more regularly than random such as corn in a cornfield.

iii) Clumped distribution (the most common in nature) in which individuals or other components are more irregular than random as for example a clump of plants arising from vegetative reproduction, a flock of bird, of people in a city.

Age distribution: The proportion of individuals of different ages in the group. Genetic characteristics especially applicable to population ecology as for example, adaptiveness reproductive fitness and persistence. Individual organisms are born, have age, grow and die but such characteristic as birth rate, death rate, density and other listed above are meaningful only at the group level. If we are to understand thoroughly the ecology of a species, we must study and measure these population group characteristics as well as know the life history and identifying features of the species. And of course, we all must be concerned about the vital statistics of human population.

Population Sizes

Some of the most important aspects of population structure are population size, density, spatial distribution age and sex rates, breeding structure and social organization. Population sizes may vary from a few individuals to millions of individuals. The population of whooping cranes (Grus American) numbers about 50 individuals a single group of birds barely clining to existence. Other rare and endangered species around the world also represent very small populations. The world's population of the Indian lion (Panthera Ieo) numbers less than 200 individuals living in the Gir forest of western India. The California condor (Gymonogyps californicus), golden marmoset (Leontideus rosalia) Javan rhinoceros (Didermocerus sumatrensis) orangutan (Pongo pygmaeus), Gray jungle fowl (Gallus sonnerati) and many other species are now represented by small remnant populations wavering on the edge between life and death. At the other extreme, the world contains many species of tremendous abundance. A population of starlings in a sigle winter flocking roost in eastern United States may number 5 million birds. The white-tailed deer population of Pennsylvania was estimated in 1946 to contain 1 million deer; the rhesus monkey population of one province in northern India was estimated in 1960 to contain between 800,000 and 1 million monkeys (it now probably contains less than half that number); the Norway rat population of Baltimore was estimated in World War II at 400,000; and the bison population of the Great Plains was estimated to contain 60 million individuals in the early nineteenth century, though by the 1890's the bison population had crashed to less than 600 through the greed and disruptive influence of man.

Population Estimates

These figures raise the question; of course as to how such data are obtained, and this is a large and complicated subject in itself. Censusing wild animal population is fraught with many difficulties of sampling error and statistical bias. Occasionally direct counts are possible as in aerial photographs of big game herds or sea bird colonies, but for most animals this is impossible and various kinds of sampling methods must be employed. Most animals are not readily visible because of their behavior and habitat, or because they exist in such abundance or scarcity that they cannot be readily counted.

It therefore becomes necessary to estimate numbers through programs of capture and resembling. With many animals systematic program of capture-mark-recapture has been helpful in obtaining estimates of abundance. The ratio of marked to unmarked animals in subsequent trapping runs provides a population estimate known as the Lincoln Index. If, for example one is studying a resident bird or rodent population in a forest, and succeeds in capturing and banading 100 individuals all of which are released, and then if on a second trapping program, one again catches 100 animals of which 20 were previously banded one would estimate the total original population at 500 animals.

The validity of the method involves several major assumptions: (1) that the marked animals mix randomly in the population, and (2) that the probability of recapture is the same for each individual regardless of whether or not it is marked, and (3) that there be no immigration or emigration;, death or births between the sampling periods. Actually, these conditions are rarely attained in natural populations. Many studies on birds and mammals have shown that some individuals are easily captured whereas others are rarely captured. There are thus "trap-happy" individuals and "trap shy" individuals.

In some species, one experience with trapping on the part of an individual animal may reduce the probability of subsequent recapture, whereas in others it may enhance it. Such behavior patterns bias the data and reduce the value of the population estimates made with this method. Also, birth, death, immigration and emigration are often continuous processes within populations, so their effect can rarely be eliminated entirely.

Nonetheless, the Lincoln Index approaches validity in certain cases. It can be a helpful tool for population estimates, particularly in the following cases: (1) if a capture method is available which is relatively nonselective and random in its action, (2) if a relatively confined population is available so that emigration and immigration do not occur, as an island population, a woodland population surrounded by forest, an oasis, etc., (3) if the population can be sampled at a time when it is relatively stable, not reproducing during the sampling period, and not migrating. Under these circumstances, capture-mark-recapture technique can produce the best available result on certain animal populations.

Another method of estimating animal population size is based on a change in sex ratio before and after a known harvest of one sex has been removed. This has been called the Kelker Ratio, and has been used in the estimation of deer, pheasants, and other wildlife populations where hunting statistics are available. In a study of California deer by Dasmann, the sex ratio observed before the hunting season was 74 bucks to 139 does, and after the hunting season in the same area it was 50 bucks to 139 does. The harvest in the hunting season was known to be 246 bucks. Hence the removal of 246 bucks population should be in the same proportion to the original sex ratio as the total buck harvest was to the change in sex ratios produced by the harvest.

Population density

Population density is an important aspect of both population size and distribution. In many populations, the precise boundaries of the population are unknown. Though the distributional limits of the species may be recognized each species usually consists of several populations. The essential feature which distinguishes a population is whether or not the individuals are potentially interactive, or potentially interbreeding.

If a species consists of two or more discrete aggregations, which do not interact or influence each other these aggregations can then be considered populations. If this condition persists over an extended period of time, different races, subspecies and even species may emerge as gene frequencies change and evolution occurs. Since populational boundaries are often unknown the most practical description of population abundance is frequently in terms of densities.

As an example many ecologic studies of small rodents have shown population densities usually in the range of 0.5 to 20 animals per acre. But showed population densities of the deer mouse (Peromyscus leucopus)in southern Michigan woodlots to range from 3.08 to 10.32 individuals per acre, Blair estimated the population density of beach mice or old field mice (Peromyscus polionotus) to range from 0.83 to 1.41 individuals per acre. Ashby estimated population densities of the bank vote (Clethrionomy glaeolus) a relative of the meadow mouse, to run as high as 17 to 18 individuals per acre in central England. An important consideration, which will be emphasized later, is that many populations fluctuate widely in density and size in the course of natural events. Occasionally small mammal populations reach amazingly high densities. In California, Krebs described a meadow mouse or vole population (Microtus Californicus) which achieved a density of 150 to 300 individuals per acre in a relatively small area, and Parson (1963) described a population of house mice in California which increased from a density of less than 5 mice per acre to 300 mice per acre within 6 months. At these high levels, or course, rodents can become agricultural problem and can precipitate economic losses of crops or stored food products.

Spatial distribution

Another important consideration in concepts of both population size and density is the spatial distribution of individuals. Spatial distribution involves a fascinating interplay of behavioral and ecologic factors, and is a prime example of the inseparable tie between behavior and ecology. We have been speaking of population densities as though animals were randomly distributed. This is rarely true, and it is much more common for population to be clumped and nonrandomly distributed. This might accurately portray the distribution of insects in an old field, isopods or millipedes in aforest, nesting birds or prairie dogs in a grassland, baboons in a savannah, fish in a lake, or even plankton in the sea.

The reason for such clumping or nonrandom distribution might be numerous. In some cases clumping would be attributable to the heterogeneity of the environment the distribution of food, cover and shelter. For example, dung beetles in an old field would be clumped around the dung isopods and millipedes around the fallen log, etc. In other cases, the clumping might be due to the social behavior. This clumping may be related to environmental resources, but not necessarily so. This is not to say that random dispersion of animals never occurs. Park pointed out that flour beetles (Tribalism confusum) are distributed in flour according to a Poison distribution, indicating that the individuals are spaced at random.

Cole found that certain spiders were distributed randomly under boards placed on the forest floor. These examples are probably exceptions to the general rule, and they are most likely attributable to either the homogeneous or artificial nature of the environment in each case. For purposes of generalization it can be stated with reasonable assurance that most animals and many plants in natural environments show a clumped or nonrandom pattern of distribution.

Sex ratios and age structures

It is obvious that animals within a population differ in a number of ways: sex, age, breeding condition health, physical condition and social status. Demography, or the numerical analysis of populations, is concerned with the sex, age and breeding condition of the population. Medical science and epistemology are primarily concerned with the health and physical condition of the population and behavioural science is primarily concerned with the social and behavioral states within the population. In most vertebrate populations primary sex ratios at hatching or birth approach 50 percent male and 50 percent female.

In some forms, there is a slight deviation from this; in man, for example sex ratio at birth is closer to 52 percent male, and this is true for rabbits cattle and many birds. In domestic chickens, sheep and horses, sex ratios at birth or hatching are usually closer to 49 percent male. The secondary sex ratios or adult sex ratios of many vertebrates show greater diviations, however. In the Alaskan fur seal one adult bull may dominate a breeding group of 30 females.

Wheras the total adult male sex ratio is greater than 3 percent the excess adult males are on the periphery of the breeding grounds, they are essentially surplus, and they suffer substa-ntially higher mortality. In most primate populations, adult females out number adult males two or three to one. Apparently males have higher mortality rates throughout life than females have higher mortality rates than males, particularly during the nesting season. In waterfowl and gallinaceous birds, for example male often outnumber females. The sex ratio of blue-winged seal (Querquedula discors) in fall and winter flocks averaged 59 percent males in a sample of 5,090 birds and in the early summer breeding population of mallards (Anas platyrhynchos) and pintail (Anas acuta) in Manitoba, males may outnumber females 4 or 5 to 1.

Age ratios represent another important element of population structure, and one that is also particularly valuable in analyzing populations dynamics Figure2 shows age structure of populations in terms of three basis types: (A) a declining population, with a low percentage of young in the population (B) a stable population with a large percentage of young than adults, and (C) an increasing population, with a very large percentage of young. The exact shape and proportions of these age pyramids are a function of natality, mortality and population turnover.

Some populations characteri-stically have tremendous production of sex cells and youngs with massive morality of young. The Pacific herring, for example produces 8,000 eggs per famale per season, of which 95 percent hatch but only 0.1 percent survive to maturity. The female shad lays from 30,000 to 100,000 eggs per female per season, and less than 0.1 percent grow into mature fishes. Populations as such would have age pyramids with very broad bases with tiny segments ascending into adult levels. At the other extreme, elephants and whales usually produce only one young per female every few years, and they would have age pyramids with relatively narrow bases and expanded adult proportions. That is, the majority of animals in the population would be adults. Rhesus populations in India show different age structures in different habitats. In roadside habitats and villages, where juvenile rhesus monkey (1 to 3 year of age, postweaned but preadult) are trapped for export for use in biomedical research and pharmaceutical production throughout the world, there is a conspicuous juvenile "gap" that is, an unnatural shortage of these individuals.

These populations have been declining throughout the 1950's and 1960's at the rate of approximately 5 percent per year, as India has exported any where from 40,000 to over 100,000 rhesus monkey per year, primarily juveniles. In temples where rhesus monkeys are protected to a certain extent for religious reasons, and in forests where they can escape trapping more successfully age structure of an island population of rhesus monkeys on Cayo Santiago Island off the eastern coast of Puerto Rico is also shown. This population has been fed and protected by a governmental research program of the National Institutes of Health, and it shows a broad base of infants and juveniles. This population was increasing at the rate of 16 percent per year in the early 1960's and gave a good indication of the natural tendency of primate populations to increase if protected in a satisfactory habitat.

Population theories

The question of population has occupied the attention of state men and philosophers since ancient times. But it is only recently that attempts have been made to investigate systematically the cause of population change and the specific ways in which population dynamics influence human welfare. Plato and Aristotle were interested in the question of population size in the context of a city-state. To them the optimum size of the population was that is with which man's potentialities could be fully developed and his "highest good" realized. Thus was possible when the population range was enough to be economically self-sufficient and capable of defending itself, but not too large for constitutional government. Plato specified 5,040 as the number of citizens "most likely to be useful to all cities". European authors on population during the early modern and the medieval periods favoured a growing population. The discovery of the New World, the increase of commerce between Europe and Asia, and the rise of national state brought about some changes in the terms of discussion of population questions, but not until the later part of the eighteenth century was there any marked shift in the general attitude favouring large and increasing population.

The mercantilist and cameralist schools of political economy, which flourished in Europe during much of the seventeenth and the eighteenth century, emphasized the economic, political and military advantages of large and growing population and favoured various measures to stimulate population growth. These writers were concerned primarily with the ways and means of increasing the wealth and power of the State. Their aim was not to promote per capita income but to increase the aggregate national income which was viewed as a sources of revenue for the State.

During the last half of the eighteenth century, a number of writers on economic and social questions rejected the mercantilist doctrine and the view that population growth was advantageous and should be actively encouraged by the State. Some writers, specially in England, France and Italy, emphasised the dependence of population upon means of subsistence. They opposed arrangement for poor relief on the ground that it might undermine frugality, make labour immobile, depress productivity and thus increase the pressure of number on means of subsistence. These arguments were turned against such advocates of social reform as Godwin and Condorect, in an effort to show that any benefits from reform would be cancelled by a consequent increase of populations.

At the close of eighteenth century, Thomas Malthus, an English economist, noted this tendency for populations, including human population to increase rapidly. In 1798, he published his famous work, Essay on the principal of population, in which he put forth the idea that populations tend to increase faster than their means of subsistence. In general, he felt that populations tended to increase geometrically or exponentially, whereas their food supplies and means of subsistence tended to increase only arithmetically.

A geometric increase is a series of numbers having a common ratio, such as 2...4...8...16...32...64 in which each succeeding number is twice the former. An arithmetic increase is a series of number having a common difference, such as 2...4...6...8...10...12, in which each number differs from the other by 2. Thus, Malthus thought that populations constantly tended to outstrip their food supplies and were then decimated by poverty, starvation, disease, warfare, or other catastrophes. Malthus was primarily concerned with human populations, and he felt that moral restraint was the solution to human population pressures.

Malthus's theory is thus based on two postulates and one assumption. His postulates are that (i) food is necessary for the existence of man, and (ii) that passion between sexes is necessary, and will remain nearly in its present form. The assumption is that the production of food increases in arithmetical progression while the population increases in geometrical progression. Thus the power of population is indefinitely greater than the power of the earth to produce subsistence for men. Efforts must, therefore, be made to keep the population from growing beyond the means of subsistence through some powerful checks, else it will lead to "vice and misery." Malthus talked of two kinds of checks to population positive and preventive. He termed preventive check as "prudence" and included in it postponement of marriage and restraint on reproduction. Positive checks he termed as "natural" as "they arise form the situation itself" and included in it wars, pestilence, epidemics, diseases, famines, natural calamities and the like.

Malthus was highly praised and also bitterly criticised for his views. His Essay aroused a storm of controversy which long out-lived Malthus himself, and which made both his followers and his opponents conscious of the need for adequate information on population trends and their influence on social and economic conditions. Thus Malthus was indirectly responsible for developing population censuses a vital statistics.

Malthus has been criticised for his postulates, his assumptions and his conclusions. While it is not to be disputed that food is necessary for man, but so is water clothes, shelter, and such other prime necessities of life. There is enough historical evidence to indicate that the output of food and of other necessities has increased faster than population, and has been brought about without any distinct symptoms of increasing "misery and vice", but rather of their diminution. With improvement in techniques of production increasing returns are being obtained from land and from the production of prime necessities of life. Malthus has proved a false prophet. While he postulated great miseries on account of unprecedented growth of population, it is found that birth rate in most of the advanced countries of the world has declined, and that the rapid growth of population in under-developed countries is basically due to a rapid fall in death rate and not on account, as Malthus feared, of an increase in birth rate. Marx rejected the Malthusion theory of population. He held that there is no universal law of population, and that the source of "over-population" was not to be found in man's biological power of reproduction, but in prevailing capitalistic mode of production. Over population arose because capital accumulated less rapidly than did the supply of labouring population was not only a necessary consequence of capita-listic accumulation, but also a necessary condition for the existence and continuation of the capitalistic form of economic organisation.

The realisation that Malthus has over-generalised a particular case lead to a re-thinking of the population question. Writers at the beginning of the present century began thinking that population growth is not always undesirable. This led to the development of the optimum theory of population which is associated with the name of Professor Cannan and others. The theory says that given the natural resources and the technique of production, that population, Cannan envisaged a point of maximum return for all industries taken together, that is the population which will maximise productivity. He also indicated that the optimum size of the population changes as circumstances change and as new techniques of production are introduced.

Effort to precisely define the "optimum" led to the refinement of the concept, but at the same time to the theory being criticised as of little practical value. It is considered highly artificial to select "average real income "as the soletest of the optimum. In fact there can be an optimum size of the population from ecological, political and other points of view. Again, the optimum size is subject to frequent changes on account of changes in techniques of production. The transition theory, development in the 30's of third century and associated with names of Warmn S. Thompson and Frank W. Notestein, explains the relationship between population with and economic development. It says that in developing countries, which are characterised by heavy pressure of population, low productivity heavy dependence on agriculture, rudimentary techniques of production, under-developing means of transport, and poor sanitary conditions, birth and death rate high.

Death rates are high because of poor diet, primitive sanitation, and the absence of preventive and curative medical facilities. Birth rates are high because of social beliefs and customs which encourage large families and which group up if a high death-rate community is to continue its existence. The birth rate around 40 per one thousand person and the death rate around 35 so that the population does not grow at a fast rate. With economic progress the death rate begins to decline. This is because of improved means of transport, better sewage disposal and improvement in drinking water facilities. But the birth rate stays high so that the difference population tend to grow at the rate of 20-30 per thousand population per year. Because of the rapid rate at which the population grows, this period is also known as that of "population explosion".

One of the features of economic development is typically increasing urbanisation and in an urban setting children are usually more of the burden and less of an asset. The processed economic change also weakens the force of traditional customs and beliefs. Females are being to realize that if burdened with a large family they are unlikely to play their due role in society. Consequently, the idea of large family yields place to the small-family-death and the birth rate declines from the high 40 to about 17 per one thousand population. The death rate is also low, about 8 per one thousand, so that the population just maintains itself. The three stages described above are also called the stages of slow population growth, rapid population growth and of stable or declining population growth. The industrially advanced countries of the world have passed through the first-two stages and are current in third stage. Developing countries of Asia, Africa and Latin America are either in the first stage or entering the second stage.

The sequence of events described by the theory can be traced in every region when the economy has evolved from an agrarian to an industrial market oriented economy. But it is not specific to quantity if the degree of decline of the vital rates. The theory, however, contains one significant generalisation, namely, the decline in birth rate occur after a substantial time-lag in comparing with the decline in death rate, and that during this lag population increases very rapidly. For instance, the "area of European settlement" increased its population six fold during the period 1750 and 1950. The population more than doubled in the century from 1750 to 1850 and nearly tripled in the period 1850 to 1950.

The theory is of considerable significance to under developed countires of Asia, Africa and Latin America. Their birth rate is high and the field of public health, it has been possible for the death rate to be substantially reduced with marked changes in the economy or in the birth rate. Consequently, their population is growing at a rate which doubles itself in twenty to twenty-five years. This is tending to retard economic progress. There is therefore an urgent need for bringing about a substantial decline in birth rates if economic development is to become a continuing process.

Logistic growth

The first major scientific challenge to Malthusian doctrine as a pattern of population growth appeared in the 1830's in the work of Pierre Verhulst. In 1839 Verhulst proposed that population normally grow in a much more orderly fashion than that proposed by Malthus, describing in their growth a curve of S-shaped proportion. This became known as the Logistic Theory of population growth. The logistic theory asserted that population have a slow initial growth rate, which increases exponentially unitl it reaches a maximum, and then becomes progressively less as the population approaches an upper limit of its growth. The upper limit is approached gradually and in an orderly, predictable manner. The resulting curve is the familiar S-shaped of population growth. It should be noted that the logistic curve and the Malthusian or irrupting curve do not differ in the early stages of population growth—both show a slow start followed by a period of exponential or geometric growth but they do differ fundamentally in the upper or controlling stages of growth.

The Malthusian curve is characterized by an erratic, often catastrophic, pattern of limiting growth, whereas the logistic curve characterized by a smooth, orderly and gradual pattern. Following the work of Verhulst, the logistic theory lay dormant unto it was independently derived and popularized by Raymond Pearl in the 1920's. Pearl applied the logistic curve to the population growth of yeast, Drosophila and man. This ushered forth a burst of activity resulting in the application of the logistic curve to the population growth of protozoa, water fleas, pond snails, thrips, ants, bees and other organisms by many investigators. With this wealth supporting evidence, the logistic thoery attained wide acceptance. It was proclaimed as a law of population growth by some biologists and was used to predict future population levels of experimental and natural population. Pearl's investigation led him to conclude: "It has been shown in what has preceded that populations of living thing `from the simplest' represented by yeast, to the highest, represented by man, grow in accordance with the logistic curve). One can now feel more certain that this curve is a first and tolerable close approximation to a real law of growth for human populations.... We can, for example, upon a more adequate scientific basis mere than guess-work, predict future population, or estimate past populations, outside the range of known census counts." That some biologist accepted Pearl's interpretation rather freely is indicated by statements of Clarke: "Population of a wide variety of organisms, ranging from bacteria to whales, have been found to follow the logistic curve in the growth from.... The growth of man's population follows a similar pattern whether examined in individual regions or in the world as a whole. "Clarke presented a graph of population growth in the United States fitted with a logistic curve, and observed an upper symptote of 184 million in year 2100 A.D.

Many other biologists and mathematicians, however, severely criticized the significance of the logistic theory. In fact, the theory met stern opposition since its first expression could be made to fit a wide variety of curves, often of very different nature. Gray, Hongben, and Wilson and Puffer registered doubt as to the universal validity of ligistic theory and commented on its inadequacies as a description. Feller pointed out that the actual fit of the logistic curve to much of the published data was poor and better agreement could be obtained with other formulas.

Andrewartha and Birch and sang accurately pointed to limitations of the logistic equation. Sang in careful reexamination of population growth in Drosophila, emphasized that, "The data we have summarized show that only in very exceptional circumstances would logistic growth occur in a Drosophila culture... The ecological situation is too complex to be adequately described as the Pear-Verhulst law, "Sang continued to discuss general aspect of the logistic theory by stating, "This acceptance of a plausible formula, in spite of its inaccuracy, may be justified; provided this attitude does not inhibit future research, particularly in to fundamental conception. In the case reviewed, it seems that wide acceptance of the logistic law has led to just this kind of inhibition of further work, and that many interesting ecological and physiological processes have been ignored as a result.

The study of vertebrate population in both natural and confined circumstances has produced some population growth curves deviating considerably for logistic growth and tending to ward errupting or Malthusian growth. The sheep population of southern Australia showeed a pattern of liogistic-like growth from 1840 to 1890, but its upper level was reached abruptly and was followed by substantial mortality. The deer of Kaibab National Forest in Arizona grew from an estimated population of 4,000 animals in 1918 after a period of extensive predator control to an estimated population of 1,00,000 in 1924 but in the winter of 1924, they died by the thousands after the population had destroyed its food supply.

Their growth and decline described a classic Malthusian curve. The pheasant population of Protection Island, Washington, showed a phenomenal increase from 8 birds on the 397 acre island in 1937 to nearly 2,000 birds by 1942. In one year the population showed an increase of 3000 percent. Unfortunately the field research on this population was terminated before natural control occurred, but the population seemed to be describing a typical irrupting population curve.

Several studies of seminatural populations of wild house mice have also shown irrupting population curves. Seminatural populations have been maintained in confined, but free-ranging conditions with a natural habit for this species, except for the element of confinement. Various factors such as space, food supply, resting cover and the structure of the physical environment have been experimentally controlled to elucidate their roles in population growth and limitation Work of this type at the University of Wiconsin and the Institute of Ecology in Warsaw Poland, has shown that population growth in *house mice* is highly variable, and often dependent upon patterns of sicial behaviour within the population. For example, populations in identical environments showed widely variable patterns of population growth. The variability was related to social behaviour within the population in terms of aggressive behaviour, dominance hierarchies, sexual behaviour and parental care.

All populations become controlled even though abundant food and nesting space was still available when aggressive behaviour increased to an abnormal degree, when the nesting territories of females were no longer properly maintained, and when parental care of the nestlings become aberrant. Infant mortality increased through nest desertion and cannibalism. The densities at which these behavioral pathologies occurred were highly variable between populations, and they were function of the crowding tolerance and behavioral stability of individuals. This complicated interplay of ecology and social behavior was not adequately accounted for by the logistic theory and the resulting population curves were totally individualistic and unpredictable.

A fairly complete evaluation of logistic theory has been presented by Andrewartha and Birch. They have analyzed the inherent assumptions of the logistic equation and have reviewed many studies in the light of these assumptions. The analysis has led them to conclude that the logistic curve provides an inperfect, yet useful, description of the population growth of organisms with simple life histories (such as yeast and Paramecium); but for animals with complex life histories (such as insects and vertebrates) the population growth usually does not conform closely to the logistic curve except for small parts of the data. The concluding attitude of Andrewartha and Birch is well experessed in the following. "despite its theoretical limitations, the logistc curve remains a useful tool for the ecologist. Because of its limitations, too much reliance should not be placed on it in particular cases until it has been verified empirically for each case. "Hence, it appears fallacious to utilize the logistic equation of a prognostic purpose in any population without experimental precedence.

The evidence is clear, therefore, to support the early contention of Wilson and Puffer that the logistic formula should not be considered a law of population growth to such an extent that it permits extrapolation of the curve of forecasting purposes or the interpretation that the constants of the equation are the constants for nature. Rather, it does provide a convenient description of growth in certain well-studied populations. The major question in human terms is, of course, what pattern of population growth characterizes man? From the early seventeenth century to the present time, the world's population has displayed exponential growth.

There seems to be widespread faith that human populations shall ultimately reveal a logistic pattern, but certainly all attempts to predict human population levels by means of the logistic model have been very far from the mark. Pearl predicted a world population of 2.6 million by the year 2100. This population level was exceeded many years ago, and economic planning based on Pearl's prediction would indeed have been disastrous.

Although we continue to hope that human population throughout the world will achieve a pattern of graduallogistic balance through orderly and intentional process, there is no sound scientific evidence that this will in fact occur. The possible of catastrophic mortality through war, disease and starvation are still very much with us.

So also are other patterns of mortality which are not necessarily catastrophic in their total effect, but are clearly related to density, crowding and modern population pressures. In this category are automobile accident mortalities, which are responsible for 50,000 deaths per year in the United States alone, various stress-related diseases, including coronary artery disease and high blood pressure, and various pollution-related diseases including emphysema and chronic bronchitis. If deaths attributable to these causes continue to show present trends, these mortality factors will begin to play a more significant role in population regulation, especially in technologically advanced countries.

Population Fluctuations

Different patterns of population growth may be exhibited by populations recovering from major depletions, or by populations invading new inhabitants. Since most populations are already established, however, and are not growing de novo, they are instead showing varying degrees of fluctuations. Our familiar concept of the balance of nature implies stability but the fact is that most animal populations under go continual change. Natural populations often exhibit a dynamic resiliency, rising and falling in response to many factors. In one sense, a pattern of population fluctuation such as that normally exhibited by most species, may be considered continuous series of growth & decline sequeuces. Thus study of population fluctuation becomes closely allied with the study of population growth and natural control.

Seasonal fluctuations

At the outset it is convenient to separate population fluctuations which are seasonal (i.e., obviously related to seasonal weather), from those which are nonseasonal latitudes, most animal pupulations have breeding seasons so that population growth typically occurs in the spring and summer seasons. Most arthropods and vertebrates begin producing young in the spring and summer seasons, so that these seasons are characterized by population growth, and they cease the production of young in laten autumn and winter. In aquatic ecosystems, many populations also show marked seasonal fluctuations. Both phytoplankton (miscoroscopic algae), and zooplankton (protozoa rotifiers, small crustacea, etc.) usually show spring and fall increases in population. These sharp increases, often 10 to 20 fold, are known as "pulses". They are sometimes correlated with temperature changes or with turnovers in the water state which recirculate nutrients.

In the tropical regions of the world, where there are not sharply defined seasons on a temperature basis, reproductive seasonality still occurs in many plants and animals, and it is often related to the cycle of wet and dry seasons. Many tropical insects, for example, have sharp peaks of abundance related to the monsoon seasons. Even vertebrate animals, including primates, have been found in nature to have definite periods of breeding and birth seasons, often correlated with the monsoons. In both tropic and temperate zones an important principle seems to be that the young appear at the most favourable time of the year in term of food and climate. Thus young deer and antelope are born when insect food is most readily available. These are exceptions to this general principle; for example, young rhesus monkeys are born in the midst of the hot dry season in northern India, when temperatures are excessively high (often over 100 degree F), water is extremely scarce, and plant growth is virtually nil. In this case, however, the infant feeds exclusively on the mother's milk for two or three months until the monsoon season starts. It is possible that natural selection has acted in such a way as to place breeding, conception and the first trimester of pregnancy at the most favorable time of the year, namely, the immediate postmonsoon month when food is most abundant.

In other words, the events of conception and early embryonic growth may have been more critical in the life cycle of the rhesus monkey in terms of evironmental influences than the period of early infancy which is well protected by the mother and the social group. This is speculation and not established fact at this time. It also fits with the observation, however, that the monsoon season with its luxuriant growth again arrives when the infant is two or three months old and is relying less on maternal milk and starts taking its own food.

Population fluctuations which are relatively independent of seasons are of two general types: (1) random, or (2) cyclic. Cyclic fluctuations are those which have a definite periodicity, or rhythmicity; that is, the population reaches a peak at fairly regular intervals. In vertebrate populations, certain species reach population peaks every 4 years, while other have a 10-year cycle. These will be discussed later as specialized cases of more generalized random fluctuations.

Ramdom fluctuations

Ramdom fluctuations may be minor perturbations of fairly stable population, producting relatively flat curves, or they may be major changes in abundance, producing erratic unstable curves. Examples of stable populations are increasingly difficult to find in the modern world. They were probably characteristic of many organisms in undisturbed ecosystems; that is, ecosystems in balance. They were also probably characteristic of complex systems, such as forests and particularly tropical forests where great species diversity produced a complicated web of natural check and balances on each species. Within such a system most vertebrate populations were probably stable and exhibited only their typical seasonal fluctuations. Probably some of the best modern examples of stable populations may be found among the birds and larger mammals. In birds where considerable species diversity occurs and fairly extensive food resources are available, considerable stability in population size may be present. John Gibb reviewed a number of bird population studies several years ago and noted: "Notwithstanding the considerable annual fluctuation is bird populations that have been recorded in almost every species investigated, it is their comparative stability over long periods that is their most remarkable attribute. "Thus, although wide variations is number occur form year to year in many bird population, these population often show a tendency to return to typical levels. An example of a relatively stable vertebrate population is afforded by a census study of rhesus monkey populations (Macaca mulatta) in northern India. With complete protection, good health and abundant food supplies, rhesus populations can increase in experimental colonies at the rate of 15 per cent per year. But in India, where monkey experience limited food supplies, varying degrees of protection from the human population, and a considerable number of infections diseases, populations do not show this pattern of increase.

In general, throughout India as in most of the world, primate population are declining due to habitat deterioration (cutting of forest, overgrazing of savannah lands, etc.) human competition (where native people shoot monkeys for food), and high rates of trapping for commercial trade. As an example of the latter, in the late 1950's India exported more than 100,000 monkeys per year for use in biomedical research and pharmaceutical production, though the current figure is only about onehalf of this. These monkeys were essential in the development of polio vaccine, measeles vaccine, and many types of medical and psychological research. The long term population trend of many animals, especially wild vertebrate animals, is down ward. Since 1600 A.D., 120 species of birds and mammals have become extinct, a number several times greater than the natural rate of extinction.

At the present time, approximately one vertebrate animal become extinct on the face of the earth every year, and within the next 30 to 50 years, 100 more species will probably be lost. The International Union of the Conservation of Nature periodically publishes a list of rare and endangered species (the publication is known as the Red Data Book), and it now lists nearly 300 species of birds and 275 species of mammals as endangered. The peregrine falcon (Falco peregrinus), ospery (Pandion haliaetus), and bald eagle (Haliaectus leuocephalus) are presently among the vanishing animals of North America. Others have come close to extinction, but were saved by intensive conservation movement at the last moment.

Throughout the world, the basic causes of extinction are habitat deterioration, poaching, poisoning, and direct or indirect competition with man. From the ecological standpoint, some extinction is an inevitable evolutionary process, but when it becomes excessive as in the last century, it is a serious disruption to biotic communities and ecosystems. Each extinction lessens species diversity, and thereby reduces ecosystem stability. Each estinction also represents the irreplaceable loss of unique biological material.

It should be pointed out of course, that some vertebrate animals have increased drastically within the last century, but these are usually domestic animals such as dogs, cats, cattle, goats and so forth, or wild animals that have learned to become commensal with man; that is, animals which have successfully adapted and even exploited the human environment. This includes the domestic rats and mice such as the Norway rat, the house mouse the Asian bandicoot rat (Bandicota bengalensis),the Asian house shrew (Suncus murina) to starling, the house sparrow and the pigeon (Columba livea.) Even the herring gull (Larus argentatus) is adapting to man's urban environment successfully by feeding on garbage dumps, and gull population on both United States coasts have increased in recent years.

Apart from those species which seem to be headed towards a permanent decline and eventual extinction, and those which thrive on man-made environments, the fact remains that most animal population fluctuate substantially from year to year. It is often a matter of common knowledge among farmers and sportsmen that "this is a good year for pheasants,"or "there were rabbits all over the place last year, but not so many this year," or "this is the world year for mosquitoes we've ever had," and so forth. These are all popular expressions of population fluctuations.

Scientifically, these fluctuations are not often recorded in a systematic and continuous fashion, but some data do exist to document the types of fluctuations which are common in animal populations. In trying to manage animal populations, we tend to do the most obvious thing, which would logically seem to produce results, and often does, but in the long run, may be quite detrimental to the total ecological balance of a community. Thus, in controlling insect populations, we use residual poisons such as DDT, but are now discovering after some 30 years, that this has long term side effects which are quite dangerous, and we are also dicovering that many insects have become resistant to it in any case. In controlling the population levels and general health of their prey species. In enhancing the abundance of desirable species, we have often poured money into artificial breeding and restocking programes of animals such as pheasant, wild turkeys and trout only to find that in the meantime, the habitat has so deteriorated through pollution that the transplants have little or no change of success, and the recovery rate per transplant is almost nil. This is not intended as a blanket condemnation of all of our control and management programs, but by and large the balance sheet is not favourable, mainly because we have based our management on trial and error rather than on sound basic knowledge of the ecological factors involved in population growth and regulation.

Population Cycles

The most diamatic population fluctuations are those which are cyclic: that is, in which the population reaches peak numbers at fairly regular intervals. Some animal populations, such as those of the snowshoe hare (Lepus americana) have periodicity of abundance of 9 to 10 years. Other animal populations have a cyclical periodicity of 3 to 4 years. These include the Norway lemming of Scandinavia (Lemmus lemmus), the field mouse (Microtus agrestis) of Europe, and some of the predators of these two notably the colored fox (Vulpes fulva) and the snowy owl (Nyctea scandiaca). Not all populations of these animals always fluctuate cyclically. When they appear to do so, sophisticated statistics may be required for proof of real cyclic periodicity.

These animal population cycles have been intriguing and puzzling to biologists for many years. Considerable research has been centered around trying to understand the cause of cycles, and numerous theories have been proposed to account for them. The final answers are still not clear, though many biological phenomena have been discovered and analyzed in this search for understanding. In general, two types of factors have been considered as the basic causes of animal population cycles: (1) extrinsic factors; that is, those outside of the population itself, such as climate and weather, cosmic events competition between species, predator-prey relationship's and food supply; or (2) intrinsic factors; that is those within the animals as population increases occur and they become crowded.

Several of these factors can be dismissed rather quickly, since extenive research has failed to show any clear relationship. For example climatic and meteorological changes have not been satisfactorily correlated with regular long-term cycles of population change in any animal. Sometimes population peaks coincide with years of great rainfall, or some other notable aspect of weather, but these occurrences are by no means consistent. At one time, sunspot cycles were considered to be correlated with population cycles. Sunspot maixma have occurred every 10 to 11 years, and they were correlated with peak population of hares and lynx in the late eighteenth and early ninteenthe century, but in the last 100 years, have been entirely out of phase.

Changes of populations of predators such as the lynx, fox and sowy owl are clearly correlated with cycles of their prey. But they cannot be shown to cause the prey cycles, for prey cycles often occur in areas devoid of predators. For example snowshoe hare populations have cycles in areas where no lynx are found. There has been some attention given to disease and parastism as the cause of long-term cycles. Twenty years ago there was interest in hypoglycemic shock in snowshoe hares.

Animals at peak periods of abundance seemed to have low blood sugar levels and go into states of shock and death. This was interpreted as the alarm reaction of Selye's general adaptions syndrome. There has also been considerable work on tuberculosis and toxoplasmosis in wild voles and although high disease rates have occasionally been asociated with die-offs of peak population, these associations have not been consistent. Many rodent declines have occourred without evidence of an infectious disease as the cause. Most ecologists have not been well trained in pathology, bacteriology and virology, not have scientists in these disciplines extended their work satisfactorily to population ecology.

Food supply has often been considered inportant in the decline of animal populations. Lack proposed that food is the main limiting factor in bird populations as it is for deer population in North America, Most field studies of rodents have demonstrated however, that even at peak periods of abundance ample food is present. Chitty studied grass during and after high vole plagues and found ample food. There are exceptions; for instance, during lemming peaks food becomes severely depleted and this is one of the factors that is though to start lemming migrations. Pitelka felt that insufficient attention has been paaid to qualitative study of food.

7 Future of Man

The most difficult and essential problems for man's future lie in broad ecologic and social issues. No two questions are more fundamental or more troublesome to the future of man than the ecologists' inquiry. "What is the carrying capacity of the earth for man? In other words, what are the limits of the life support capabilities of the earth in terms of human population?" are the behaviorists' concern. "What are the tolerance levels of man to the pressures generated by his own societies? In other words, if man is his own worst enemy, does this lead to his self destruction?"

The carrying capacity, or life support capability, of the earth for man may be dependent upon a great variety of factors. We can readily identify oxygen water, food space and shelter as vital requirements for human life. There might be many others as well such as mineral resources, energy availability, or the concentration of toxic metabolites, which could reach states above human tolerance levels. Most of these however, would operate through the media of air, water or food.

Oxygen supply and Air Quality

The atmosphere of the earth consists of about 78 percent nitrogen, 21 percent oxygen, 0.03 percent carbon dioxide, and the remainder, other gases. This composition is remarkably stable despite the fact that oxygen is a highly reactive element which readily combines with a great many other elements in chemical processes of oxidation. The ecologist Lamont C Cole (1970) has pointed out that oxygen has remained stable in the earth's atmosphere because green plants recycle it in molecular form through photosynthetic process. Approximately 70 percent of the free oxygen produced each year is through to come from planktonic plants in the ocean, and the remaining 30 percent from terrestrial vegetation of forests and grasslands.

Cole emphasized that if we should seriously pollute the sea, as is already happening in coastal areas, and if we should excessively denude the forests and grasslands of the earth, we shall be in danger of upsetting the oxygen balance in the atmosphere. In other words, we must look upon the oceans, forests and grasslands, as the world's oxygen tank, and we must insure their continued healthy ecologic function. This point of view is contested by W.S.Broecker, a geochemist at Columbia University Dr. Broecker points out that each square meter of the earth's surface is convered by 60,000 moles of oxygen gas. Plants in the ocean and land produce only about 8 moles per square meter per year, an this is almost entirely consumed by their own respiration and that of animals and bacteria.

Thus other organisms use all but 0.0004 percent of the oxygen produced by plants. Broecker maintains, therefore, that man does not depend upon the oxygen generated by plants, and that there is no current danger at all to his oxygen supply. "In conclusion it can be stated with some confidence that the molecular oxygen supply in the atmosphere and the broad expanse of open ocean are not threatened by man's activities in the foreseeable future. Molecular oxygen is one resource that is virtually unlimited". Broecker feels that man's existence will be threatened by other types of environmental pollution long before his oxygen supply becomes endangered, and that we need not worry about total oxygen supply. This is typical of some of the different opinions which are unfolding between physical and biological scientists. Such differences often indicate the need for more research and better knowledge before we can fully accept one position on another.

Another atmospheric problem of long-term concern is the increase in carbon dioxide in the atmosphere. Between 1860 and 1960 the combustion of fuel increased the CO_2 content of the air by 14 percent of its formerly stable level, Carbon dioxide tends to reflect infrared rays from the earth, and this provides a "greenhouse effect" for the whole earth; that is, it accumulates heat. This will gradually increase the temperature of the earth—an effect which may have long-term implications in climate modification. A rise of just 2 or 3 degrees in annual mean temperature would have a major melting effect on glaciers and ice caps, and this could raise the level of the seas several feet.

Such a rise in the seal level would inundate many of the world's major cities. Some experts predict it may require 4,000 years, and others feel it will not happen at all. The latter believe that any occurrence of the greenhouse effect due to increasing CO_2 in the atmosphere will be offset by increasing particulate pollution in the air and a cooling of the earth's surface. Once gain, we see fundamental controversies on major issues-largely because we do not really have the complete answer. The atmospheric problem may also become acute in terms of various toxic elements released into surface, air and water: carbon monoxide, sulfur dioxide, lead from automobile exhaust, hydrocarbons from industrial operations and numerous other products of modern technological societies. For example, 500 million pounds of lead are discharged into the atmosphere of the United States each year from automobile exhausts alone, and significant amounts of this become incorporated into living organisms.

A recent study in Sweden showed concentrations of lead in mosses to be as high as 300 to 500 ppm along roadsides. Some roadside plants in the United States may have lead levels up to 3,000 ppm,. This lead enters the plants via both air and water and can then enter man and animals. A hundred years ago, before the automobile, concentrations in roadside plants were less than 20 ppm. Lead levels over 0.5 ppm have detectable physiological effects in man and animals, and 1.0 ppm levels are often toxic.

There seems to be little or no immediate threat that we will run out of air for breathing, but several warnings have appeared regarding the quality of our air. We apparently have an adequate supply of total oxygen for many centuries to come though even this is a debatable point. In some localities the situation may be critical at certain times. If stationary inversions of air trap the polluted atmosphere over some of our largest cities such as Tokyo, London, New York or Los Angels, toxic air pollutants increase to dangerous levels.

In August of 1969, a massive still-air inversion occurred in eastern United States, causing serious air pollution in urban areas from New Orleans and St. Louis, north to Chicago and Milwaukee, and east to Washington and Baltimore. At no time was there an absolute shortage of oxygen, but dangerous level of carbon monoxide, sulfur dioxide, hydrocarbons and other air pollutants accumulated in several cities. Twenty million people were affected by varying degrees of eye irritation and respiratory diseases.

Fortunately, this great blanket of smog was dispersed by upper air currents in the first week of September, and no large-scale health tragedies occurred. More serious event of this type are likely to occur in urban areas unless strong measures against air pollution are undertaken.

Water Supplies

Worldwide water supplies represent a concern for the future of man in terms of both the quality and absolute quantity of fresh water. Many cities have reached or are coming close to their total available fresh water supply and are reaching further and further to obtain new supplies. Los Angeles, for example, is planning tremendous canals and canduits which will bring water from the pacific northwest over 1,000 miles to the growing megalopolis of southern California.

Two Russian geographers and hydrologists, Kalinin and Bykov, have pointed out (1969) that "The world's water resources from a single entity. The basic hydrologic cycle of the world depends upon solar energy, atmospheric circulation, heat reflectivity and surface retention. The latter is strongly influenced by the biotic community, kalinin and Bykov point out, on the assumption that the moisture content of the earth remains constant that the total gains and losses from the land and the sea are in balance.

Thus, the total precipitation equals the total evaporative loss from land and sea. However, the relative retention time and thus the total amount of water in terrestrial and fresh-water environments can change substantially in relation to the biota and terrestrial storage capacity. A large proportion of the world's fresh water, falls and resides in relatively unpopulated portions of the globe, particularly Siberia and northern Canada.

Here, man's activities are limited by other factors and he cannot readily obtain these vast stores of water. Secondly, the matter of water quality is of immediate concern. The principle of water reuse is now becoming accepted, so that people recognize much of their water has been used previously by other municipal systems.

Each reuse cycle however, is increasingly costly in terms of treatment, and each reuse cycle requires disinfectants and other industrial chemicals, combining these problems of distribution and quality, it seems that the water problem of the future is not going to be one of a global water shortage on a planetary basis but one of local crises due to the concentrations of people, industry and water use.

Certainly, the immediate concern is finding local water resources of adequate quality to meet the burgeoning needs of urban populations. To the city dweller in New York or Hong Kong whose water tap runs dry or to the refugee in the Middle East who is dying of thirst, it well be little comfort to realise that 29,000,000 cubic kilometers of fresh water are locked in glaciers and polar ice caps.

World Food Supplies

Nowhere is scientific controversy more evident than on the issue of world food supplies. Some scientists predict famines and mass starvation within the next 10 years. They feel the stage is irrevocably set for millions of people to die of starvation in the 1980's. In fact, we already have famine in many parts of the world, if one defines famine as people dying of lack of food. Even at the height of the Green Revolution in Asia in which new strains of rice and wheat were producing phenomenally large yields, famine conditions existed in parts of Rajasthan in India: in sections of Burma and in southern Thailand.

Starvation is now a grim reality in much of the world and it may well get worse despite the Green revolution. Other scientists refute the pessimism and find evidence for a hopeful outlook in the world food situation. They feel that the Green Revolution which was based on new strains of rice and wheat in Asia and Africa, and which produced a doubling and tripling of yields in one or two years, shows what can be done with modern agriculture. For the first time in many decades, agricultural output increased more rapidly than population growth. India began to approach self-sufficiency in food production, and the Philippines began to export rice again for the first time since 1900.

A word of caution about the Green revolution is also necessary. It is quite true that new strains of agricultural crops can sometimes dramatically increase yields, but often these new strains require more careful management, more intensive care, more water, more fertilizer and more insecticides. They may be more vulnerable to plant viruses or fungi.

Although the plant geneticists try to breed in every possible disease resistance, these new strains do not have the centuries of natural selection and regional adaptation behind them as do the local strains which the villages have always used. In West Bengal and Orissa, there were reports that some of the newly introduced high-yield strains of rice were succumbing to a disease that was not affecting that old traditional strains. If this happens to a major extent, the Green Revolution could collapse and a severe blow would be dealt to the villager's faith in the agricultural scientist.

Another factor in the world food picture is one of pure climatic good fortune. Green Revolution in India was accompanied by the best monsoon seasons in over 10 years. A long series of droughts had broken is southern Asia and adequate rains fell in the right pattern. The possibility exists, therefore, that a good part of the success of the Green Revolution was due to good weather, combined with new strains and more fertilizer. This is a question that will be more adequately answered in future year to see if increasing agricultural production can be maintained in years of poor rainfall.

In India too much rain came and it fell to early. The wheat crops in northern India were magnificent in April and May just before harvesting and threshing, but heavy premature rains in many areas caused the wheat to rot before it could be threshed. Bihar had serious crop losses from flooding. Once again, this was a tragic and dramatic demonstration that even the Green Revolution remains dependent on climatic factors outside the control of man. Through all of these arguments on how many people the world can feed, one must ask the realistic question: "Is it necessary that there always be a race between population and food production? Is it desirable to be teetering on the brink of starvation at all times? Is it the best representation of the human condition to be struggling in this manner to feed more and more people at a subsistence level?

One also has the impression in reviewing the literature on world food supplies that too much emphasis is placed on future predictions, and not enough on current conditions. We furthermore know that malnutrition exist throughout the world, even in the most advanced countries. We furthermore know that malnutrition of many kind has profound effects upon growth, development and behavior.

Children growing up with malnutrition from early infancy definitely show reduced growth rates and possible retardation of physical and mental development. They furthermore show increased susceptibility to infectious disease. Thus, we return to the issue of the quality of life, and the realization that merely preventing death by starvation does not by any means insure a quality nutritional base for adequate growth and human development. It is apparent that the problem of world food supplies will continue for some time as a problem of both quantity, and quality and especially one of distribution.

Space on Earth

In 1958 Paul Sears wrote an important essay entitled, "The Inexorable Problem of Space," in which he concluded that "Our future security may depend less upon priority in exploring outer space that upon our wisdom in managing the space in which we live". The earth's total surface is 196,940,400 square miles of which about 30 per cent or 57,230,000 square miles is land surface. Of this total land surface, approximately 5,000,000 square miles are in Antarctica deserts, and 11,000,000 square miles are in mountainous regions.

Of deserts and moutains, at least one-fourth of their jano area cannot be inhabited, so this eliminates another 5,000,000 square miles of land surface for human habitation. We are left with approximately 47,000,000 square miles of land surface, at the most, for the world's population. The world now has nearly about 6 billion people. The world population has doubled in the past 40 years and may double again in the next century, perhaps approaching stability at about 11 billion by the year 2100. By 1990, of the world's 5.3 billion people, 4.1 billion—77 percent lived in the developing world; 1.2 billion inhabited the industrialized countries.

India has nearly 1/6th of the world's population. There are 257 person per sq. km. in India. In the Capital City Delhi there are 6318 person per sq. km. Thus it is to say human population is not evenly distributed. It leads to the problem of space. As with water, air and food, it is important to distinguish between total or potential amount of any given resource on earth and that amount which is actually utilized or practically available to man, it was pointed out in the sections on water and air that the earth's total amount is apparently adequate for many years to come, but the supply in any given city may not be of sufficient quantity or quality to support the standard of living we would like.

The amount of living space per person in some cities is less than 20 square feet per person. Theoretically, we could redistribute these crowded people to provide more even spacing around the world. Obviously, this cannot be done, and if it were attempted it might wreak havoc with agricultural production. The economic and behavioral needs of man are such that he must cluster together, but in so doing, he creates superdensities such as Calcutta, London, New York, Paris, Tokyo, Shanghai and Buenos Aires. The slums of which are definitely deleterious to human health and welfare. Why, we might ask, does man insist on this degree of crowding and then complain about it so bitterly? Why do people continue to pour into the cities of the world when they are already known as centers of population crime and illness?

Perhaps we need to know more about the real spatial and social needs of man. What are his requirements for space and privacy, and how do these interact with his needs for social stimulation and group activity? Does man become conditioned to crowding in early childhood and then seek further crowding throughout his life? Does he require the social pressure of a crowd because he feels a compulsion to be in touch with the group and thus be able to predict what the group is going to do next? Have we produced, in our urban centers, self-destructive, systems which so modify the behavior of their inhabitants that these inhabitants exaggerate their own problems and can no longer achieve rational solutions? These are all tough question that cannot be answered in simple or easy terms. The answers will require the best intelligence we can muster from many disciplines. Some sorts of answers will certainly be necessary before we can even begin to solve the ecologic and behavioral problems associated with our own use of terrestrial space.

Distributional Inequality

With all of the last four life-support requirements of man-air, water, food and space—we have seen that the major problem of immediate importance lie in distribution more than in total amount. Technologist therefore look upon these problems as soluble by technological means. They feel, they are essentially matters of economics and transportation.

Behavioral and environmental scientists see them as more basic, however, relating to man's fundamental ecologic behavior. They feel, they are essentially matters of economics and transportation. Behavioral and environmental scientists see them as more basic, however, relating to man's fundamental ecologic behavior. They feel that certain basic social and ecologic tendencies of man prevent equitable distributions of the earth's resources. Obviously this type of argument can go on forever, and it is perhaps necessary to take a pragmatic approach and look at the current picture for the best glimpse of the future. Many of the industrial nations of the world, especially the United States, the national northern Europe and Japan, have enjoyed rising prosperity for the last 25 years.

In ecologic time, of course, this is but a flash in the pan, but let us assume that it can be sustained for another 25 or 50 years. It is worth examining where the material and energetic support for this prosperity arises. This distributional inequality is another major symptom of ecologic instability in the world. One would hope for some narrowing in the gap between the haves and the have-nots, but unfortunately the opposite trend is still occurring. The wealthy nations of the world are far outstripping the poor nations in their rates of economic gain and human betterment.

There are, however, self-corrective mechanisms within this system as, for example, when some segment of the labour it can find which will do a satisfactory job. Thus, there has been a great expansion of manufacturing in Japan, Hong Kong, Taiwan and Singapore in industries as diverse as steel, electronics, textiles, toys and optics. The Western factory worker experiences for the first time an improving situation. The western industrial worker continues to refuse his own working prospects by striking for higher and higher wages. This will certainly be one mechanism by which wealth is more adequately balanced between nations.

Still the mass of poverty throughout much of Asia, Africa and Latin America is so great that it is difficult to imagine a painless equilibrium. The differential distribution of wealth remains sharp and represents a type of ecologic situation which cannot last indefinitely without severe consequences and reverberations.

To the ecologist it means either a successional stage which is not yet mature, or a system headed for a major breakdown. Economic and ecologic maturity should ideally be in the direction of broader and more equitable distribution of the earth's resources—a narrowing of the gap between the haves and the have-nots, both within and between nations, provided the total system does not pass beyond the state of corrective change.

Violence and War

Any ecological system or biotic community can be overloaded to the point of collapse by a variety of factors. Certainly one of the factors for man of most immediate and realistic danger is that of social violence. Either by war or by civil riot, man threatens his own existence. To maintain a sense of physical security man has built an incredible stockpile of destructive power. The nations of the world spend more than 200 billion dollars per year on armaments—more than on health education, welfare, agriculture, child care and care for the aged combined. Twenty tons of explosives are stored in the world for every man, women and child. Russia and the United States have stockpiled 6,000 nuclear war heads possessing thousands of megatons of explosive force. Each megaton has more destructive power than which existed in any of the previous wars of history. The world spends almost 8,000 dollars per year training each soldier and less than 100 dollars per year educating each child.

The question is, of course, how long can man continue to preside over armament like this without using it? And the answer seems to be, not very long. The fact is we do use part of the power in an attempt to achieve social goals. It is being used daily in various parts of the world to maintain territory and dominance ranks. It is justified as a spare and deliberate use in hopes of avoiding the all out effort. How long this can continue is anyone's guess, and the ecologist can claim no more omniscience than the political scientist.

The ecologist points out, however, that the ecologic and social pressures which generate international military tension may rise with increased population pressure. It is not likely that man can solve the problems of war and violence until he approaches and understands their basic orgins in terms of the behavioral and ecologic relationship between social groups, and the environment.

Adaptive Capabilities of Man

Many scholars feel that one of the great problems of the present and the future is the rapid rate of change which is occurring in society and technology. There is not, however, agreement on the consequences of the rapid change, changing life patterns are too rapid for man to adapt successfully. Mankind is so overstimulated by the increase in new sight and sounds, so bombarded by world wide mass media, and so constantly beset with noval situations and new problems that he will break down psychologically and physically. His responses will be maladaptive and pathologic. He will thus collapse under sensory overload. It is quite possible that each of these points of view has large elements of truth in it. In the present world one can find many examples of both. Our mental institutions are already filled with those who could not adapt to sensory overload and who experience a personal breakdown of physical and mental health. Yet one can find people are living a sub-human existence where the quality of life is so low that the amount of human pain and misery seem unbearable. Yet man has adapted successfully to these inhuman conditions.

In many ways, the future is with us now. We can find throughout the world an abundance of high quality life in high quality environments, but we can also find an abundance of low quality life in impoverished environments. Many of the questions about the future revolve, therefore, around the question of relative balance and proportions. Which way are we heading and which way will the scales tip? A major aspect of the future centers around our sense of responsibility to posterity. No representative of posterity is here to speak for the segment of society. How do the present populations of mankind regard their obligations to future generations? Are we willing to be more conservative in our life requirements, and are we, in fact, willing to reduce our rates of personal economic growth to insure a better environment and a greater abundance of resources for future generations? This may be one of the central issues of the next twenty years, and one on which the quality of present and future environments depends.

World-Wide Environmental Issues

Environmental problems have become a global issue: Some of the world wide issues are briefly described below:

(a) Climatic changes/Green house effect

Many studies show that the atmospheric pollution is on a global scale. Today, however, it is not only the air over our cities that is polluted, the entire atmosphere of our planet is now afflicted to some degree. Meteorologists talk about a nebulous veil of air pollution encircling the entire earth. Smong has been observed over the North Pole, and in their unlikely places. There are range of environmental pollution issues originating largely in developed countries which are also of global significance. Carbon dioxide (CO_2) build-up leading to climate change is one of those key issues. Recent research indicates that increasing concentrations of CO_2 in the atmosphere, due mainly to the burning of fossil fuels could cause a warming trend leading to climate change in the next century. While there is still scientific uncertainty about the pace and impact of such a change it could be of sufficient magnitude to produce major physical, economic and social dislocations on a global scale.

A special challenge posed by the issue is that if preventive actions are postponed until some effects are perceived, it may be too late to avoid appreciable climate change. It is therefore in the interests of all nations to support international efforts to understand the phenomenon at early enough stage for action to be taken to reduce or avoid major economic, social and environmental disruptions around the world.

However, as developed countries are currently responsible for over two thirds of total world emissions of CO_2 by fossil fuel combustion, they have a special responsibility for supporting and contributing to accelerated efforts to assess more precisely the nature of the problems and the alternatives for action to reduce or avoid any serious risks. Another issue is the world-wide distribution and use of harmful or potentially harmful chemicals. There are now some 70,000 chemicals on the commercial market, and many are currently used and released into the environment with little or no knowledge of their potential long-term effects. It is estimated that another 1,000 new chemicals enter the market each year. Chemicals cross national boundaries in a variety of ways: they are transported by winds and water.

(b) Risks to Ozone Layer

Ozone is a different form of life sustaining oxygen, having three atoms instead of the usual two. But this third atom, scientists have found, is very unstable and can be eliminated by chlorine ions. The scientist have now found a close link between the family foam mattress, refrigerator and the air conditioning system of an average home and Ozone layer, 50 to 70 kilometers above the earth. They believe that free chlorine ions into the atmosphere, from plastics and other chemicals remove that extra atom of oxygen which makes Ozone. Only, in this case, the loss of the one atom changes the ability of the layer to act as a strainer to absorb the damaging ultraviolet B wavelengths from the sunlight.

The findings by the experts have revealed that the rate of Ozone Layer depletion is sufficiently large to present serious consequences for man, animals and plants. In addition, new threats to the ozone layer are being recognized. Take, for example, the refrigeration industry. Food preservation has become a necessity and refrigeration is here to stay. Only advanced technology can provide the solution to prevent the release of the damaging ions. Another example is nitrogenous fertilizer which, it is suspected, release chemicals damaging to the Ozone layer. The demand for more food and the booming green revolution in many developing countries have pushed up the use of this fertilizer. Today the world uses about 50 megatons of it. In another 20 years the rate of uses will be 100-200 megatons.

Though the Ozone Layer accounts for less than one millionth part of thc earth's atmosphere, it is of vital importance to the earth's life form. It has two main functions: first it absorbs the UV-B while allowing the heat generating infra-red rays to reach the earth freely. The heat from infrared rays sustains plants and human lives by keeping the temperature of the planet at a desired level. Secondly, it influences the whether by stabilizing the stratosphere which acts as a cap to the turbulent weather system in the troposphere.

Scientists knowledge of the Ozone layer is still limited for want of advanced technology. The presence of other substances in the stratosphere, which absorbs UV-B, makes it difficult to come to a numerical and logical conclusion. But the scientists do agree that an average 70 per cent of the total UV-B reaching this planet falls on the equatorial regions between the latitudes 30° S and 30° N. They also agree that for genetic reasons people with light eyes and light skin (including albinos) are most affected by UV-B.

Scientists today claim UV-B can destroy micro-organisms outright as well as cells in plants and animals. It also destroys the DN molecules and the genetic cell composition. Under natural conditions plants and animals have developed behavioural adaptation but gradual or sudden increase in UV radiation due to changes in the Ozone Layer is likely to disturb this balance. While the scientists agree that UV-B contributes to the induction of cancer they agree the dose response relationships are too complex and unknown. These make the quantitative analysis speculative, as regards the cause of melanoma cancer and resulting deaths. The North European Conference of Scientists November, 1980, warned that non-melanoma skin cancer in light-skinned people would increase and many crops and many acquatic organisms, including shrimps ad fish eggs, could be damaged by increased UV-B infiltration. This could even affect the reproductive process of plants and animals.

It recommended that existing satellite and ground based ozone observing system should be integrated through Global Ozone Observing System (GOOS). They also recommended updating of equipment and setting up of more stations for observing the ozone system. The scientists concerned with outer limits of the biosphere felt they are not yet technically equipped to measure accurately the rate of ozone depletion-already depleted by one percent. Though the danger is known, science has not yet provided a perfect model of measuring the depletion rate.

(c) Acid Rain

Every one realises how important the rain is. Without it the continents would be barren. Yet now the rain in parts of the earth has taken on a new and threatening complexity. It mixes in the air with pollution from burning fossil fuels particularly in power stations, factories and motor vehicles and brings down dilute sulphuric acid. This is killing fish and water life, and corroding buildings, including some of the world's most important ancient monuments. It may also damage forest and croplands, and possibly pose a substantial threat to health. Acid rain is now recognized as an international problem. The air of industrial towns partly is cleared by chimneys at factories which push the pollution high into the air. These chimneys have made the things better locally by dispersing the pollutants, but aggravated the international difficulties. For the sulphur and nitrogen compounds emitted by burning fossil fuels can be blown thousands of kilometers by the winds, to cause acid rain in countries far from their points of origin. A special conference on acidification of the environment was held in Stockholm in 1982. It reviewed and assessed a large amount of scientific information not previously available.

So far, acid rain has been seen as an essentially regional problem, confined to the industrial areas of the northern hemisphere. But, though the problem was first perceived there, it may become much more widespread for acid rain is likely to occur wherever fossil fuels are intensively used. Acidification is an environmental problem, or becoming one, in parts of Europe and North America. Similarly polluted areas are likely to exist elsewhere in the world, especially around large urban and industrial coagulumerations. Soils are normally much better able to resist acidification than lakes, rivers and streams. Meanwhile, acid accelerates corrosion in most materials used in construction of historic buildings, and other important cultural objects. Some of th worlds greatest cultural treasures including Parthenon in Athens and Trojans' column in Rome are being eaten away by acid fall out.

There must be more effort to research and develop ways of improving energy conservation, environmentally appropriate technologies for producing heat and power, and techniques for removing sulphur from fossil fuels and gaseous emissions. The problem of acidification may well spread to new areas as a result of rapid industrialization and the growth of cities in other parts of the world, particularly developing countries. So it is important that areas affected by acid deposition and susceptible to damage from acidification are identified as soon as possible. It they are, the damage could be mitigated or even avoided of a minimum cost to society by initiating research and applying what we already know about pollution control.

(d) Pollution of World's Oceans

The data collected from the Baltic, the North Sea, and some North American and Australian stuaries and coastal waters suggest that in most of these areas the pollution that caused most concern was due to sewage, agricultural chemicals, oil and metals. Metal concentrations were clearly elevated in coastal waters, and in fish and shelfish living there. In some areas, mercury levels in species such as tuna were high enough to make these fish unsuitable as human food. Some of the pollution came via river; the amount of iron, manganese, copper, zinc, lead, tin and antimony that reached the sea by this route was far greater than would be supplied by natural geological processes. Other contaminants came through atmospheric deposition: the importance of this pathway for metals and synthetic chemicals was increasingly recognized during the decade.

Offshore oil and gas exploration and dredging for sand and gravel in coastal areas increased during the period 1972-1982. Coastal zone development affected extensive estuarine areas, as well as mangrove swamps and coral reefs. Oil pollution killed sea-birds fouled beaches and affected tourism. Although tanker accidents were the source of less than 5 percent of all the oil entering the sea, accidents released large volumes in small areas, and were therefore especially damaging.

A recent report issued by a group of experts on marine pollution and published by the United Nations Environmental Programme, Warns that increasing contamination of the coastal seas, which has so far only caused localized disturbances to the marine environment may become much more serious. The combined effects of the dumping of sewage sludge industrial wastes, litter and oil spills into the Mediterranean, the North Sea, the Baltic and the Gulf of Mexico, may well cause ecological disasters which could gradually lead to a global crisis since the oceans are a unified life support system. The report points out that the sea's capacity to absorb wastes has its limits, so that as the world economy continues to grow, even more sewage, industrial wastes and other effluents are likely to find their way into the sea.

Man-made radioactive materials have been discovered in the open ocean and scientists have found traces of DDT in th Arctic sea and the Antarctic penguin, thousands of miles away from the areas where the pesticide is used. And because many of the disturbances which are introduced occur at vulnerable points such as the surface of the sea or on the sea floor, where oceanic life-supporting processes are generated, the dangers and magnified. Although the modification of this microlayer has occurred only on the vicinity of oil spills and around the marine dump site of municipal wastes and no evidence has yet been found that detergents in the open ocean have significantly influenced the properties of the sea surface, the report points out that environmental catastrophies in certain coastal areas and seas might already have occurred.

Only improved management of the coastal areas and seas which involves controlling more closely the dumping of pollutants and careful monitoring of contaminant levels, will prevent a further deterioration of coastal environments. Much work however is still to be done before the several ecological sub-systems which constitute the ocean are fully understood. By the end of 1982 action plans were in operation or planned for 10 regions involving some 100 coastal States. Coastal and marine national parks and reserves were extended.

(e) Desertification

Desertification is now perceived as an intensifing world wide threat which, to varying degress, affects over 100 Member States of the United Nations. More than 20 percent of the world's population (850 million people) live in the 35 percent of the Earth's land surface (45 million km) which consists of the arid, semi-arid and sub-humid zones which is at risk of desertification. Three quarters of this area is already at least moderately desertified. Each year an additional 21 million hectares of agricultural land deteriorates through desertification to a point at which it is no longer economically productive. The last production has been valued at 26 billion dollars annually. Although desertfication is exacerbate by severe drought, its principal cause is human over exploitation of dry lands through over-cultivation, overgrazing, poor irrigation practices and deforestation. When drought strikes the overtaxed livelihood systems, desertification results.

The most devastating environmental and economic effects of desertification are in developing countries in the drylands. National balance of payments problems and local rural poverty lead to over-exploitation of lands and forests for food, fuel, cash crops and meat for export. As these and other pressures increase to unsustainable levels, agricultural productivity and soil fertility decline, creating in turn new pressures which lead to further degradation. Recovery of land for agricultural purposes become costly and often economically unfeasible, thereby reducing the economic potential of the affected countries. In many of these countries agricultural land is also being reduced by salinization, urban sprawl and sand-dune encroachment.

Increased desertification in one part of the world can indirectly induce desertification else where, as loss of agricultural land and increased food demand raise Prices and intensify pressures on marginal drylands in other food exporting and importing countries alike. There are also other impacts on international trade, to the extent that developing countries affected by rapid desertification will have a reduced capacity to import goods and services from developed countries and from other developing countries.

Other globally shared impacts include loss of valuable genetic resources, disruption of hydrological cycles and increase of atmospheric dust. A great deal is already known about the causes and effects of desertification and about the measures needed to halt or combat it. Moreover, there is favourable cost benefit ration to justify, preventive action.

The 1980 estimate for implementing the 1977 plan of Action to combat Desertification was 2.4 billion dollar annually over 20 years, which is only a fraction of the estimated value of agricultural production lost each year through desertification. It is, therefore, in the interests of all countries to establish some concreate cooperative actions between groups of developed and developing countries to halt further desertification, a goal which can be achieved, given a sufficient allocation of resources, on the basis of currently available knowledge.

(f) Loss of Tropical Forests

Although tropical forests are found in early half of all United Nations Member States, most are in twenty countries and nearly half their total area in three countries alone (Brazil, Indonesia and Zaire). The tropical forests are the world's richest biological zones, and are estimated to contain as much as 40 per cent of all terrestrial species on the planet.

In addition, tropical forests provide a wide range of useful products of economic significance for both developing and developed countries. Undisturbed tropical forests are also home to millions of the world's tribal peoples. Many experts contend that tropical forests are being exploited at a rate and in a way that is ecologically destructive and economically unsustainable. There are varying estimates of how much of the tropical forests will be lost by the year 2000 through complete conversion to other uses and by severe degradation.

The results of the tropical forests resources assessment project conducted by FAO and UNEP indicated an estimated minimum loss of closed tropical forests of 12.5 per cent by the year 2000 (an average of 7.5 million hectares annually between 1980-2000) through complete conversion to other uses alone.

Other studies which not only consider complete conversion but also include severe degradation caused, for example, by logging operations, by over-grazing, by over-exploitation for fuelwood, or by natural causes such as fires, insects and diseases, have estimated losses of tropical forests as upto 20 million hectares or more in a year. The FAO/UNEP study estimates that tree planting on industrial and non-industrial plantations accounts for only 10 per cent of the area of tropical forests cleared annually.

The major causes of deforestation differ by region. According to the FAO/UNEP report in Latin America the two major causes are shifting cultivation and the conversion to pasture land for cattle ranching. In Asia and Africa, shifting cultivation by subsistence farmers and removals for fuelwood are major causes. The extraction of timber from tropical forests for industrial purposes is not considered by the FAO/UNEP reports as a major factor in deforestation.

On a world-wide basis, the UNEP report estimates that fuelwood and charcoal removals from tropical forests are as much as eight times greater than for industrial wood. Serious degradation of loss of watershed forests in particular has a wide range of major ecological and economic effects through increased erosion, floods, landslides, and silting of hydro-electric facilities, irrigation systems reservoirs and harbours. The lives and livelihood of half the world's population depend directly on the wise management of watershed forests and eco-systems.

Deforestation also influences local, regicnal and possibly even global climatic conditions by altering hydrological regions, reducing vegetation cover and causing more heat and light to be reflected back to the atmosphere (the albedo effect), and through accelerated burning of the forest biomass, which contributes to carbon dioxide build-up in the atmosphere and the global warming trend. The economic, social and political implications of possible regional and global climate changes have only begun to be assessed.

(g) Soil Loss

Almost every environmental concern connects clearly with the importance of soil. Soil acts as a reservoir for water and other inorganic substances cycling through the eco-system. It provides habitat for many of the different populations or decomposer organisms that control the rate and capacity of nutrient recycling. The importance of soil to agriculture can hardly be over emphasised. Agriculture provides about 44 percent of the poorest countries gross domestic product and 83 percent of their employment. Growing more food, building irrigation projects, conserving nature and precious wildlife and other needs must depend on proper soil management.

The wide-ranging importance of soil confirms that soil management must consequently be many-sided. The broad range of natural and man-made stress that causes the two type of soil damage: physical displacement and alterations in soild structure and composition. Uncontrolled shifting cultivation, salinization and water-logging flooding and plant cover destruction are same. An appropriate use of chemicals, weapons testing and deployment in warfare are some other casuses of soil nutrients and other types of soil degradation.

Soil damage is also caused by patterns of dry land mountain side cultivation that carried out without arrangements for bunding and terracing to protect the topsoil. Continuous cropping to meet growing food needs leave title time for soil conservation techniques. Deterioration of the productive capacity of the soil is most acute in areas affected by erosion, drought, mineral stress and other related factors. Soil resources that are vulnerable to these factors constitute 82 percent lands in South Asia.

While erosion is an integral part of the natural system, occurring even when land is covered by grass, or trees, its extent and intensity can be minimized provided land denudation is correspondingly reduced. The fact that it takes nature 100 years or more to build one inch of top soil should have engineered the vigilant observance of soil conservation measures. Instead, problems of soil erosion have continued ticking. Countries in East and South Asia are particularly burdened with soil erosion occuring in watersheds.

Anticipated acceleration of agricultural production and other activities between now and 2000 is casting its shadows: existing cultivable land will be further subjected to intensified use as cultivation of unsuitable areas persists; logging, firewood gathering and grazing will accelerate the elimination of vegetative cover, while the construction of roads, mine and water resources projects will be further pursued in the course of development. Such prospects place the least developed countries in an even more precarious situation. Slash and burn farming is a common enough culprit. Wielding destruction in gradual succession, it initially causes the loss of fertility by the thin topsoil, and later eroding its stability. In India, slash and burn farming is particularly rampant in the north-eastern and central regions affecting between six and eight million hectares of land.

Salinization affects 80 million hectares (8 percent) of total land in South Asia. Something should also be said about agricultural techniques which are designed to raise land productivity but instead lead to soil loss. The use of cheap nitrogen, for example, replaces nitrogens, fixing legumes, resulting in loss to topsoil. The commonality of problems, in addition to the implications transboundary soil management points up to the fact that soil problems cannot always be resolved within the context or national concerns. Clearly, what are required are mutual policies and strategies concerning land management, and the World Soils Policy provides an appropriate starting point.

Working in collaboration with the relevant UN bodies, UNEP has proposed the World soils Policy programme which endeavours to enhance international awareness, assist countries in the formulation of national soil policies, help develop technical and scientific knowledge and collect, compile and disseminate data on the use and management of the worlds soil resources.

(h) Depletion of genetic resources

25,000 species and more than 1,000 vertebrate species and sub-species are now estimated to be threatened with extinction, yet these two categories of organisms represent only a small proportion of the terrestrial species on the planet. According to some estimates as much as from half a million to a million species would be extinguished over the next two decades. Extinction of species on this compressed scale would be without precedent in human history.

As much as two-thirds of all terrestrial species and the majority of endoangered species are located in developing countries. A significant loss of species in developing countries would have a world wide effect on future agriculture, industry and science, and on human health and welfare generally. The most serious threat to many species is the disruption and destruction of their habitat by man, especially in the tropical forests which alone harbour as much as 40 per cent of all terrestrial species.

Significant species loss could directly affect human health and welfare locally, and even nationally, in many developing countries, where wild plants and animals are a source of fodder, traditional medicines and income especially for the poor in rural areas. In some countries tourism largely on wildlife is a major source of foreign exchange. In many developed countries over one-thrid of the gain in the productivity of many major agricultural crops over the last few decades has been attributed to breeding programmes using genetic material from related wild plant varieties, many of which are found in developing countries.

Moreover, many of the major crops in developed countries now have a limited genetic base. For example, more than half the Canadian prairie wheatlands are devoted to a single variety, Neepawa, which was developed with germplasm from Kenya; 72 per cent of United states potato production depends on only four varieties; and the entire United States soyabean industry is derived from a mere six plants from Asia. A significant loss of relavant wild species could limit the possibilities for maintaining or increasing the yields of major crops, such a wheat, maize, rice potatoes and for combating new strains of diseases or pests which threaten them. Genetic stocks from one part of the world frequently make an enormous contribution to another. Current research shows that enormous opportunities lie ahead, such as the utilization of Asian blue-green algae of fix nitrogen for crop fertilization, which may constitute a gain of hundreds of millions of dolars a year to the economics of developed countries alone.

Major medicines and other pharmaceutical products in developed countries are also often dependent on plant and animal species. For the United States of America it has been estimated that more than 40 per cent of the prescriptions written each year contain a drug of natural origin-from higher plants (25 per cent), microbes (13 per cent) or animals (3 per cent) as the sole active ingredient or as one of the main ones. It has been estimated that among the 10 percent of species that might be extinguished by the year 2000 there could be many plants with potential for use in new plantbased drugs and pharmaceuticals with a combined world wide commercial sales value of as much as 200 million pounds annually.

Biomedical research in developed countries is also vitally dependent on wild species, many of which are threatend with extinction. The innovative capacity and future of the rapidly expanding field of bio-technology, practised largely in developed countries, relies considerably on genetic patterns and resources from many wild plants and animals as only about 10 per cent of the world's plant species and 1 per cent of the animal, species have so far been subjected to even a preliminary screening for potential medical or commercial applications, a significant loss of species could also represent a permanent loss of sources for more effective chemicals, medicines and other useful products. In spite of this interdependent relationship, species loss continues to accelerate. Developing countries often lack the capacity and economic incentive for extensive species protection. They have derived little or no economic benefit from the development and use of genetic materials in developed countries, where the scientific and technical expertise to exploit species and their genetic resources predominantly resides. Nevertheless, developing countries are and will remain the principal source of genetic materials from wild plants and animals. This clearly shows that there is a common incentive for developed and developing nations to prevent a potentially catastrophic loss of species. The mutual advantage of comprehensive joint effort could result in new medicines and other useful products that would contribute to improved levels of human health and welfare in all countries.

Human progress has always depended on our technical ingenuity and a capacity for cooperative action. These qualities have often been used constructively to achieve development and environmental progress: in air and water pollution control, for example, and in increasing the efficiency of material and energy use. Many countries have increased food production and reduced population growth rates. Some technological advances, particularly in medicine, have been widely shared.

Failures to manage the environment and to sustain development threaten to overwhelm all countries. Environment and development are not separate challenges: they are inexorably linked. Development cannot subsist upon a deteriorating environmental resource base: the environment cannot be protected when growth leaves out of account the costs of environmental destruction. These problems cannot be treated separately by fragmented institutions and policies. They are linked in a complex system of cause and effect.

First, environmental stresses are linked one to another. For example, deforestation by increasing run off, accelerates soil erosion and siltation of rivers and lakes. Air pollution and acidification play their part in killing forests and lakes. Such links mean that several different problems must be tackled simultaneously. And success in one area, such as forest protection, can improve chances of success in another ares, such as soil conservation.

Second, environmental stresses and patterns of economic development are linked one to another. Thus agricultural policies may lie at the root of land, water, and forest degradation. Energy policies are associated with the global greenhouse effect, with acidification, and with deforestation for fuelwood in many developing nations. At these stresses all threaten economic development. Thus economics and ecology must be completely integrated in decision-making and law making processes not just to protect the environment, but also to protect and promote development. Economy is not just about the production of wealth, and ecology is not just about the protection of nature: they are both equally relevant for improving the lot of humankind.

Third, environmental and economic problems are linked to many social and political factors. For example, the rapid population growth that has so profound an impact on the environment and on development in may regions is driven partly by such factors as the status of women in society and other cultural values. Also, environmental stress and uneven development can increase social tensions. It could be argued that the distribution of power and influence within society lies at the heart of most environment and development challenges. Hence new approaches must involve programmes of social development, particularly to improve the position of women in society, to protect vulnerable groups and to promote local participation in decision making.

The systemic features operate not merely within but also between nations. National boundaries have become so porous that traditional distinctions between matters of local, national, and international significance have become blurred. Ecosystems do not respect national boundaries. Water pollution moves through shared rivers, lakes, and seas. The atmosphere carries air pollution over vast distances.

Major accidents—particularly those at nuclear reactors or at plants or warehouses containing toxic materials—can have widespread regional effects. Many environment-economy links also operate globally. For instance, the highly subsidized, incentive driven agriculture of industrialized market economies generates surpluses that depress prices and erode the viability of the often neglected agriculture of developing countries. Soils and other environmental resources suffer in both systems.

INDEX